THE MATH TUTOR'S FRIEND

complex numbers

word problems

conics

change of base

primes

base

BOOK 4

Algebra III

by Joe Cyr

Cyr, Joseph

The Math Tutor's Friend / Book 4 – Algebre III / by Joe Cyr

Third Edition ISBN: 978-1-967039-45-6 Joseph Cyr

Second Edition ISBN: 979-8-9879759-4-7 Joseph Cyr

First Edition ISBN: 9798473326406 Amazon Kindle, 2022

3 4 5 6 7 8 9

Subjects: Main category of book – *algebra – prime numbers; number bases; conics; complex numbers; dimension analysis; word problems; high-school math*

2.Other category - math tutor reference

Cover illustrations by Ramon E. Owen. Cover Design by Joe Cyr

Book design by Joe Cyr

The Math Tutor's Friend
Series

Books

1 - Pre-algebra

2 - Algebra I

3 - Algebra II

4 - Algebra III

5 - Probability & Statistics

6 - Geometry

7 - Trigonometry

8 - Pre-calculus

9 - Series index & "FLASHMATH" handouts/reference cards

THE MATH TUTOR'S FRIEND

BOOK 4

ALGEBRA III

Sections

About the "Math Tutor's Friend" Series:

Teachers impart knowledge guided by lesson plans in a course syllabus. Math tutors, on the other hand, face varying topics at differing levels in successive sessions; for example, pre-algebra, advanced algebra, geometry, trigonometry, statistics, or even calculus! This series addresses the issue and is intended to provide the tutor with needed quick review of high-school math topics. The material is not presented rigorously nor does it abound with proofs and explanation that are available in textbooks - the tutor already knows this stuff - but a quick refresher is always a good move before diving into a tutoring session.

Too many in high-school have reached the 11th and 12th grades lacking basic math skills such as knowing the multiplication table (relying on a calculator to multiply 6 by 7, for example), handling signed numbers, and performing operations involving fractions - knowledge that should have been acquired by the fifth grade! I try to address this deficiency by devoting the first few tutoring sessions to these fundamentals and providing handouts of multiplication tables,etc.

Books 1 through 8 of this series include a number of quiz questions (answers explained) that the math tutor may find useful.

Book 9 is an overall index of topics. It contains a set of summary "FLASHMATH" pages for quick reference or for use as handouts.

This third edition corrects typos found in previous editions and expands discussion of some of the topics.

Happy Tutoring!

Joe Cyr

Sarasota, FL

●

THE MATH TUTOR'S FRIEND
1101001
797
95_{10}
It's easy !!
$7AF9_{16}$
Mr. Math
121_4
Primes & Number Bases

About Book 4: "Algebra III"

Section 1, *Primes & Number Bases.*
Emphasis -:

- Prime numbers
- Factoring integers completely
- Change from a higher to a lower base

Section 2, *Conic Sections.*
Emphasis -:

- What are conic sections?
- Nomenclature

Section 3, *Complex Numbers.*
Emphasis -:

- Real numbers vs imaginary numbers
- Combining complex numbers

Section 4, *Dimension Analysis.*
Emphasis -:

- Commensurable quantities
- Base units
- Dimensional homogeneity
- Factor-label method

Section 5, *Word Problems.*
Emphasis -:

- Approach to solving problems
- Exact vs approximate solutions
- Percent problems
- Rate problems
- Mixture problems
- "Job" problems
- Apply dimension analysis

Section 6, *Interesting Math Concepts.*
Emphasis -:

- Angle measurements
- Geometric mean
- Mathematical palindromes
- Fibonacci Sequence
- Pascal's Triangle
- Euler's Formula

• •

1. PRIMES & NUMBER BASES
TABLE OF CONTENTS

page

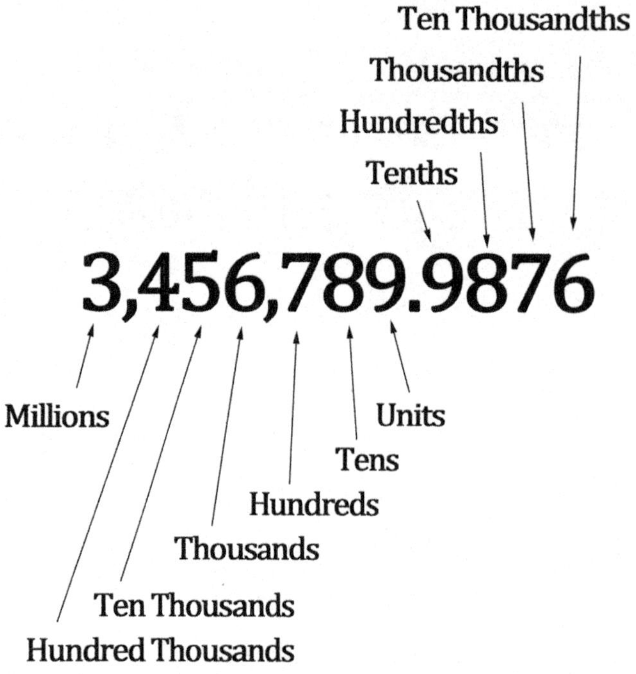

Reading a Decimal Number

DIVISIBILITY RULES*

A number is divisible by … if

2	the number is even (Note: Decimals are neither odd not even)
3	the sum of the digits is divisible by three
4	the last two digits are divisible by 4
5	the number ends in 5 or 0
6	the number is even AND divisible by three
7	9 times the units digit subtracted from the remaining digits of the number is divisible by 7.
8	the last three digits are divisible by 8
9	the sum of the digits is divisible by 9

Test number (divisible by all of the above): 362,880

* There are alternate divisibility rules.

1. PRIMES AND NUMBER BASES

Prime numbers

What is a *prime number*? Informally, a natural number (remember the definition?) is prime if and only if it is divisible only by itself and by 1. Thus, 2, 3, 7, 11, and 97 are prime numbers. There are an infinite number of primes and researchers seek the largest prime. Discussion of the largest prime* is beyond the scope of this book series.

Table of prime numbers less than 500:

2, 3, 5, 7, 11, 13, 17, 19, 23, 29, 31, 37, 41, 43, 47, 53, 59, 61, 67, 71, 73, 79, 83, 89, 97, 101, 103, 107, 109, 113, 127, 131, 137, 139, 149, 151, 157, 163, 167, 173, 179, 181, 191, 193, 197, 199, 211, 223, 227, 229, 233, 239, 241, 251, 257, 263, 269, 271, 277, 281, 283, 293, 307, 311, 313, 317, 331, 337, 347, 349, 353, 359, 367, 373, 379, 383, 389, 397, 401, 409, 419, 421, 431, 433, 439, 443, 449, 457, 461, 463, 467, 479, 487, 491, 499.

In this list of primes, note that, except for "2," they all end in (that is, the *ones*, or *units*, digit is) 1, 3, 7, or 9. Why is that?

When is a number not a prime?

Given a number, can you positively state that it is NOT prime (i.e., a *composite* number)? For example, 34,569 is not a prime number because it is divisible by 3 (the sum of the digits is 27, a number divisible by 3).

If the *ones* digit is 0, 2, 4, 6, or 8, then the number is even (divisible by 2) and (except 2, of course) cannot be prime. If the *ones* digit is 5, then the number is divisible by 5 and so cannot be prime. Therefore, if a number is prime, then its *ones* digit must be 1, 3, 7, or 9. The converse is not necessarily true.

Factoring an integer completely

You may have encountered the *fundamental theorem of arithmetic* in high school. It states, essentially, that every positive integer either is a prime number itself or can be represented as the product of prime numbers. The process to do this is called "completely factoring an integer."

Example: Let's try a complete factorization of the integer 3294 (See the divisibility rules on the facing page). First, note that 3294 is even so is divisible by 2, yielding 1647; that is,

$$3294 = 2 \cdot 1647.$$

Now 1647 is not even but its digits sum to 18 which is divisible by 3, so 1647 is divisible by 3, yielding 549. So far we have $3294 = 2 \cdot 3 \cdot 549$.

Continuing, 549 is also divisible by 3, so we now have

$$3294 = 2 \cdot 3 \cdot 3 \cdot 183.$$

* The largest known prime number (as of 6/6/2025) is $2^{136,279,841} - 1$, a number which has 41,024,320 digits when written in the decimal system.

The number 183 is also divisible by 3, and we have

$$3294 = 2 \cdot 3 \cdot 3 \cdot 3 \cdot 61.$$

From the list of primes on page 1 we see that 61 is prime and so cannot be factored further (except as the product of 61 and 1), and so we have finished completely factoring the integer 3294. What was accomplished in this exercise? According to fundamental arithmetic theorem, the integer 3294 can be expressed as the product of prime factors, and these are 2, 3, 3, 3, and 61, as we have just determined.

```
2|3,294
 3|1,647
  3|549
   3|183
      61
```

An easy way to find prime factors by hand is shown to the right; note how the successive divisions described earlier are displayed.

Creating a number that is divisible by 6

I cannot think of a single practical application for the following example, except that it is a good exercise in applying the divisibility rules.

Example: Write a 10-digit number divisible by 6.

Recall that a number is divisible by 6 if it is divisible by 2 and also divisible by 3; a number is divisible by 2 if it is even, and is divisible by 3 if the sum of its digits is divisible by 3. First, write any sequence of 8 digits. Add these digits to get a number, say n. Next, select two more digits, m and k, to add to n so that *(1)* $n + m + k$ is divisible by 3 and *(2)* k is zero or *even* (so k must be 0, 2, 4, 6, or 8).

For the example problem first write an arbitrary eight-digit number, say

$$23\ 500\ 367\ .$$

The sum of the digits in this number is 26; that is,

$$n = 2 + 3 + 5 + 0 + 0 + 3 + 6 + 7 = 26.$$

The next number greater than 26 whose sum of digits is divisible by 3 is 27, so that $m + k$ must add to 1, and k must be even (or zero) for the number to be divisible by 2. This is an easy one: $1 + 0 = 1$, so that the complete 10-digit number divisible by 6 is 2 350 036 710.

You can make the problem seem complicated by specifying other conditions, such as "the number must contain a sequence of five zeroes."

Example: 4,000,005,984 is divisible by 6. Of course, there are many answers: 5,984,400,000 would work, too.

Incidentally, a number such as 2,350,036,710 may be expressed without commas, e.g., "23 500 367 710," because in some countries the comma "," is used as a decimal point (e.g., what we know as 10.6 would be written as "10,6") so that it may be better to use a space instead of a comma to express large numbers. There are areas of the world where even the period "." is employed instead of a comma or space to separate groups of digits in large numbers.

What does the base of a number mean?

The *base* (or *radix*) is the number of different digits or combination of digits and letters that a system of counting uses to represent numbers. A common base used today is the *decimal* system ("dec" means 10); it uses the 10 digits from 0 to 9. The *binary* system uses the two digits 0 and 1; the *octal* system uses the eight digits 0 to 7; the *hexadecimal* system uses the sixteen digits 0, 1, 2, 3, 4, 5, 6, 7, 8, 9, A, B, C, D, E, F. Here are some examples of converting non-decimal base numbers to decimal.

Table method for changing a decimal to a number with base less than 10

Example: Change the decimal number 23,501 into a base-4 number.

Line	*Base* (*b*)	N_{10}	*R*	*n*	$m=b^n$	*Check:* $R\times m$
1	4	23,501				
2	4	5,875	1	0	1	1
3	4	1,468	3	1	4	12
4	4	367	0	2	16	0
5	4	91	3	3	64	192
6	4	22	3	4	256	768
7	4	5	2	5	1,024	2,048
8	4	1	1	6	4,096	4,096
9		0	1	7	16,384	16,384
					$N_4 = 11233031$	
					Total	23,501

The top row of the table holds the column titles:

"*Line*" is the step of the process.

"*Base*" is the desired base, b. Here we want to convert a base 10 number N_{10} to a base 4 number, N_4.

"N_{10}" is the base-ten number to be converted.

"*R*" is the Remainder after dividing the number in the "N_{10}" column by the new base (here, 4).

"*n*" is the power of "b," the base; begin with 0 in line 2 and increases by 1 in each step (line of the table).

"$m = b^n$" is the value of the base (4) raised to the n th power.

"*Check:* $R\times m$" is the product of the digit in the "*R*" column and the "$m = b^n$" column.

The bottom row contains the sum of the numbers in the "*Check:* $R\times m$" column, and should be equal to the original N_{10}. The next-to-bottom row is explained below. Now let's build the entire table step by step.

Example: Convert the decimal $N_{10} = 23{,}501$ to base 4 (N_4).

Line	*Base (b)*	N_{10}	*R*	*n*	$m=b^n$	*Check: R×m*
1	4	23,501				
2	4	5,875	1	0	1	1

First, insert the base, b (here, 4), in the "Base (b)" column in lines 1 and 2, as shown above, and insert the base ten number (here 23,501) to be converted in line 1 of the "N_{10}" column. Now, do the following calculations:

(1) Divide 23,501 in the "N_{10}" column by the new base, 4, and place the quotient, 5,875, in line 2 of the N_{10} column. Continuing in line 2, place the remainder (here 1), in the "*R*" column as shown. Insert 0 in the "n" column and calculate $4^0 = 1$ in the "$m = b^n$" column and $1\times1 = 1$ in the "*Check R×m*" column as shown.

(2) Increase n by 1; repeat to get the values in line 3 as shown below:

Line	*Base (b)*	N_{10}	*R*	*n*	$m=b^n$	*Check: R×m*
1	4	23,501				
2	4	5,875	1	0	1	1
3	4	1,468	3	1	4	12

Continue, increasing n by 1, until you get "0" in the "N_{10}" column, as shown below:

Line	*Base (b)*	N_{10}	*R*	*n*	$m=b^n$	*Check: R×m*
1	4	23,501				
2	4	5,875	1	0	1	1
3	4	1,468	3	1	4	12
4	4	367	0	2	16	0
5	4	91	3	3	64	192
6	4	22	3	4	256	768
7	4	5	2	5	1,024	2,048
8	4	1	1	6	4,096	4,096
9		0	1	7	16,384	16,384
					Total	23,501

To check for accuracy, sum the numbers in the "*Check: R×m*" column to arrive at the original N_{10} value, 23,501. Finally, the result we seek, N_4, is the sequence of numbers in the "*R*" column read from the bottom up.

The result is N_4 = 11 233 031. In lines 2 through 8 of the table we have N_4 equivalents of the numbers in the previous line's N_{10} column (read from row 9 up); for example, $467_{10} = 11\ 233_4$. Here's the list:

$$5{,}875_{10} = 1\ 123\ 303_4$$
$$1{,}468_{10} = 112\ 330_4$$
$$367_{10} = 11\ 233_4$$
$$91_{10} = 1\ 123_4$$
$$22_{10} = 112_4$$
$$5_{10} = 11_4$$
$$1_{10} = 1_4$$

Note that the decimal number $23{,}501_{10}$ is constructed as follows (begin with the *ones* digit and move left):

$$(1\times10^0) + (0\times10^1) + (5\times10^2) + (3\times10^3) + (2\times10^4)$$
$$= 1 + 0 + 500 + 3{,}000 + 20{,}000 = 23{,}501$$

Similarly, the base-four number $11\ 233\ 031_4$ is constructed in base-ten as follows:

$$(1\times4^0) + (3\times4^1) + (0\times4^2) + (3\times4^3) + (3\times4^4) + (2\times4^5) + (1\times4^6) + (1\times4^7)$$
$$= 1 + 12 + 0 + 192 + 768 + 2{,}048 + 4{,}096 + 16{,}384 = 23{,}501_{10}$$

Note that these are the numbers in the "*Check:* $R\times m$" column of the table. This is how a base-four number is converted to a decimal number.

We wrap up this discussion with another example: convert a hexadecimal number to a decimal number. As described earlier, the hexadecimal system uses the sixteen digits 0, 1, 2, 3, 4, 5, 6, 7, 8, 9, A, B, C, D, E, F. Here, the letters A through F are the 11th through 16th digits in the hexadecimal system.

Example: Convert the hexadecimal number $3FA2_{16}$ to its equivalent decimal. ***Begin with the right-most digit*** and proceed to each number on the left (note that A is decimal 10, and F is decimal 15):

$$3FA2_{16} = (2\times16^0) + (A\times16^1) + (F\times16^2) + (3\times16^3)$$
$$= (2\times16^0) + (10\times16^1) + (15\times16^2) + (3\times16^3)$$
$$= 2 + 160 + 3{,}840 + 12{,}288 = 16{,}290_{10}$$

Example: Convert the binary string 1101001 to its equivalent decimal number.

$$1101001_2 = (1\times2^0) + (0\times2^1) + (0\times2^2) + (1\times2^3) + (0\times2^4) + (1\times2^5) + (1\times2^6)$$
$$= 1 + 0 + 0 + 8 + 0 + 32 + 64 = 105_{10}$$ (Yep! Everything is 0s and 1s)

That's all there is to it! Remember to begin from the right-most digit. Notice that we have not discussed changing from a smaller base, say 4, to a larger base, say 11. This is beyond the scope of this book as you would need to know the multiplication table in both bases; we have learned only the decimal multiplication table.

•

THE MATH TUTOR'S FRIEND
parabola
latus rectum
hyperbola
It's easy !!
eccentricity
directrix
ellipse
Mr. Math
CONIC SECTIONS

2. CONIC SECTIONS

TABLE OF CONTENTS

•

2. CONIC SECTIONS

Conics

Conic sections, or *conics*, are formed by the intersection of a plane at some angle relative to the base of a *double-napped right circular cone.*

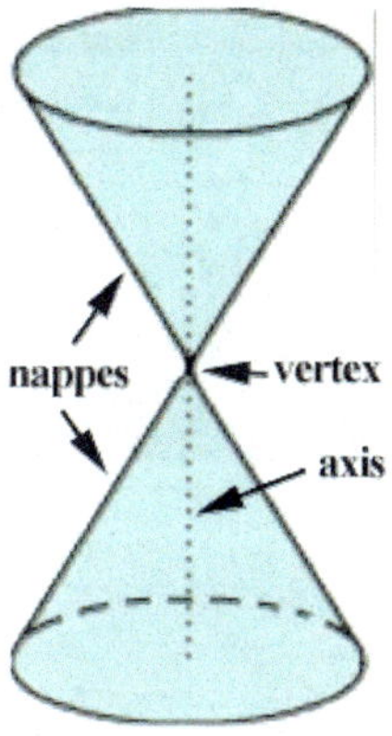

double-napped right circular cones

There are four *conic sections* as shown below.*

General form for conic equations

The general form of a conic equation is the second degree equation

$$Ax^2 + Bxy + Cy^2 + Dx + Ey + F = 0$$

where all coefficients and the constant are real numbers and not all zero. The second term, Bxy, is related to rotation and is beyond the scope of this book, so we shall ignore it and set B = 0 so that the general-form equation becomes

$$Ax^2 + Cy^2 + Dx + Ey + F = 0 . \qquad (2\text{-}1)$$

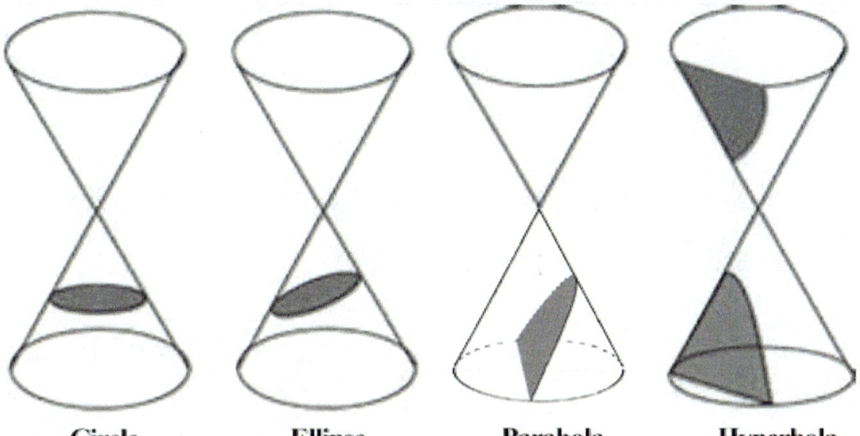

Identifying conic sections

Given an equation of degree 2 with two variables, collect *like* terms and move all terms to the left side of the " = " sign leaving zero to the right of the sign to achieve the pattern of Equation (2-1) where some terms may be missing (one or more, but not all, of the coefficients may be zero) and the variables x or y (or both) may be squared. Either variable could be replaced by a *term*; e.g., instead of $x^2 + y^2 = 1$, it might be $(x - 2)^2 + (y + 1)^2 = 1$ [D, E, and F are zero]. Now refer to the *conic-section decision tree* in Figure 2-1 on the next page.

* Also, passing a plane perpendicular to the axis through the vertex defines a point, a plane tangent to the surface ot the cones defines a line; and the plane holding the axis defines two lines. These are the so-called degenerate conic sectrions.

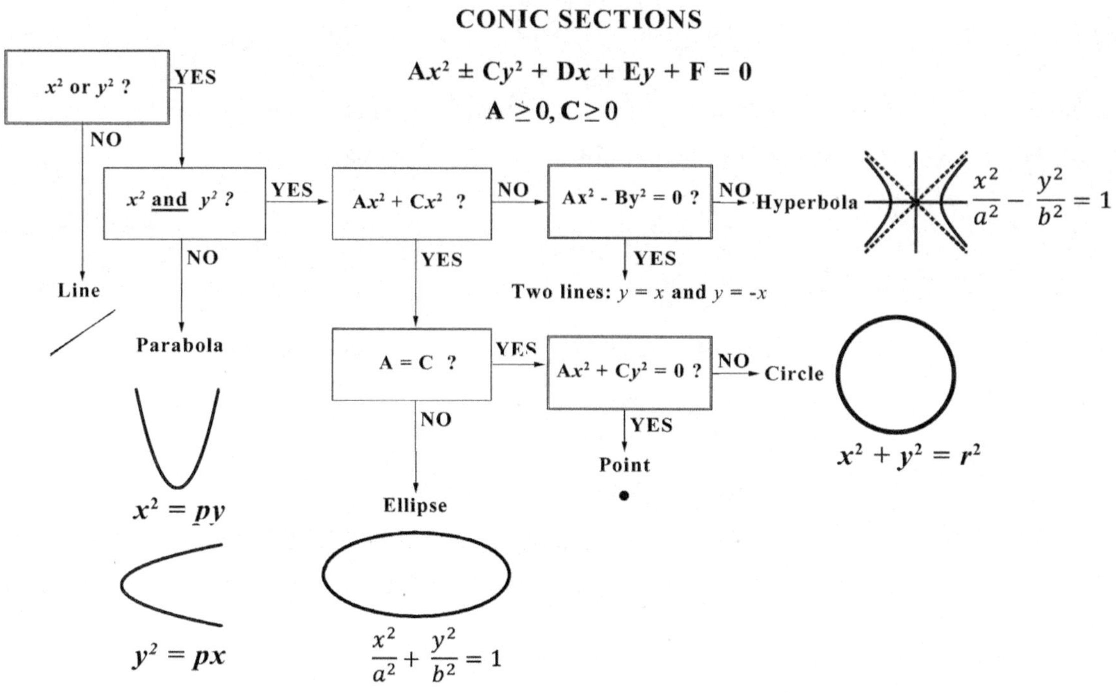

Figure 2-1

Line

If neither x nor y are squared then Equation (1) is linear.

Example: $x + y - 2 = 0$

Parabola

Both variables are present, but one is squared and the other is linear.

Example: $x^2 = 4y$

Circle

Both squared variables are present and have the same positive coefficient ($A = C > 0$). The right hand side is positive.

Example: $x^2 + y^2 = 1$.

If the right hand side is zero, it is a point (0, 0) at the origin. If the right hand side is negative, then there is no graph because the sum of two squares cannot be negative.

Ellipse

Both squared variables are present, both are positive, but they have different coefficients ($A \neq C$). The right hand side is positive.

Example: $3x^2 + 4y^2 = 1$.

If the right hand side is zero instead of 1, the expression is a point (0, 0).

Example: $3x^2 + y^2 = 0$.

If the right hand side is negative, then there is no graph because the sum of two squares cannot be negative.

Hyperbola

Both squared variables are present; A or C is positive and the other is negative. The right hand side is positive.

Example: $3x^2 - 4y^2 = 1$

Special (degenerate) cases

If $Ax^2 + Cy^2 = 0$, the result is a single point at the origin.
If $Ax^2 - Cy^2 = 0$ and $A = C$, the result is the difference of two squares: $(x - y)(x + y)$ which are the two lines $y = x$ and $y = -x$.

Discriminant

Conics can be classified from the general equation *discriminant*, $B^2 - 4AC$: Recall the discriminant of the quadratic equation, $ax^2 + bx + c = 0$ in the quadratic formula

$$x = \frac{-b \pm \sqrt{b^2 - 4ac}}{2a}.$$

There is a discriminant for the basic equation

$$Ax^2 + Bxy + Cy^2 + Dx + Ey + F = 0.$$

The familiar $B^2 - 4AC$. It can be used to identify the conic sections as follows:

Parabola:	$B^2 - 4AC = 0$	$AC = 0$ (A = 0 or C = 0 but not both)
Ellipse:	$B^2 - 4AC < 0$	$B \neq 0$ or $A \neq C$ $AC > 0$ (A & C have like signs)
Circle:	$B^2 - 4AC < 0$	$B = 0$ and $A = C$
Hyperbola:	$B^2 - 4AC > 0$	$AC < 0$ (Rectangular* hyperbola : $A + C = 0$)

Development of equations for conic sections

Let's begin with a simple line! See Figure 2-2 below.

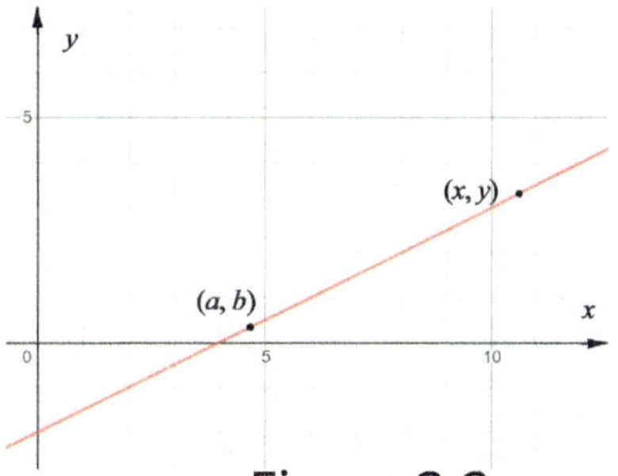

Figure 2-2

Given a line and a point P(a, b) on that line, the perpendicular distance, c, to that line from any other point P(x, y) is given by the *distance formula*

$$c = \sqrt{(x-a)^2 + (y-b)^2} \qquad (2\text{-}2)$$

* Both asymptotes are perpendicular to each other.

Square both sides of Equation (2-2):

$$c^2 = (x - a)^2 + (y - b)^2 \qquad (2\text{-}3)$$

Equation (2-3) is the equation of a *circle* of radius c centered at P(a, b).

Circle

Circle

Circle: Basic equation: $Ax^2 + Bxy + Cy^2 + Dx + Ey + F = 0$, where A, B, and C are not all equal to zero. The "B" term involves rotation of the axes and is not considered here, so we set B = 0:

$$Ax^2 + Cy^2 + Dx + Ey + F = 0.$$

If A = C then it is the equation of a circle, a special case of the ellipse where foci are merged to become the circle center. The circle is a conic section formed by a plane parallel to that of the base of the cone intersecting that cone. A circle is defined as the set (locus) of points that are equidistant from a point called the center. This distance is the radius. Algebraically, a circle is the locus of points (x, y) at some fixed distance r (the "radius") from some fixed point (h, k) called the "center." More simply, the conics form of the equation of a circle with center at (h, k) is

$$(x - h)^2 + (y - k)^2 = r^2$$

Note that this is really the distance formula for the distance of any point from the center, (h, k), of the circle. Also note that in order to determine if a given point (x_1, y_1) is inside or outside a circle one need only use the distance formula to determine if the given point is a distance less than or greater than the radius of that circle.

Example 1: Find the coordinates of the center of the circle

$$x^2 - 25 + (y - 4)^2 = 0.$$

What is the diameter?

- Put the equation into the form of the circle equation

$$(x - h)^2 + (y - k)^2 = r^2$$

and determine h and k.

$$(x - 0)^2 + (y - 4)^2 = 5^2$$

This is the equation of a circle with radius 5 centered at (0, 4). The diameter, d, of a circle is twice the radius, r; that is, $d = 2r$. The diameter of the circle is 10.

Example 2: What geometric form does the following represent?

$$4(x^2 + y^2) - 8(3x + y) = -4$$

- First, note that you can divide through by 4 to get

$$x^2 + y^2 - 2(3x + y) = -1$$

Clear the parentheses and put into the form $x^2 + y^2 + Dx + Ey + F = 0$

$$x^2 + y^2 - 6x - 2y + 1 = 0$$

Put the equation into the form of the circle equation $(x - h)^2 + (y - k)^2 = r^2$ then identify h and k. We'll need to complete the square for the "x" terms (the last three terms on the left-hand side are already a perfect square):

$$x^2 - 6x + y^2 - 2y + 1 = 0 \text{ [take half the coefficient jof } x \text{ and square it to get 9]}$$

$$x^2 - 6x + 9 + y^2 - 2y + 1 = 9 \text{ [add 9 to both sides of the equation]}$$

$$(x - 3)^2 + (y - 1)^2 = 3^2 \text{ [express as perfect-square terms]}$$

Success! This is a circle with radius 3 and center at (3, 1).

Example 3: Find the equation of a circle with center at (-4, 1) passing through point (4, 7).

• First, find the radius such that the circle passes through the given point; that is, "plug" 4 and 7 for x and y in the equation:

$$\sqrt{(4-(-4))^2 + (7-1)^2}$$

$$\sqrt{8^2 + 6^2} = \sqrt{64 + 36}$$

and square both sides; the radius is 10, the positive square root of 100, so an equation of this circle is

$$10^2 = (x + 4)^2 + (y - 1)^2$$

The *interior of a circle* consists of all the points inside the circle's line. Algebraically, the distance of an interior point from the center is less than the value of the radius. The *exterior of a circle* are all the points outside the circle's line, having distances from the center greater than the radius.

CIRCLE	
$Ax^2 + Cy^2 + Dx + Ey + F = 0$	
Conic equation	$(x - h)^2 + (y - k)^2 = r^2$
Center	(h, k)
Radius	2
Eccentricity	$e = 0$

Ellipse

Introduction: A circle is "the set of all points in a plane equidistant from a fixed point (center)." The standard form for a circle, with center at the origin is $x^2 + y^2 = r^2$, where r is the radius of the circle. See Figure 2-3.

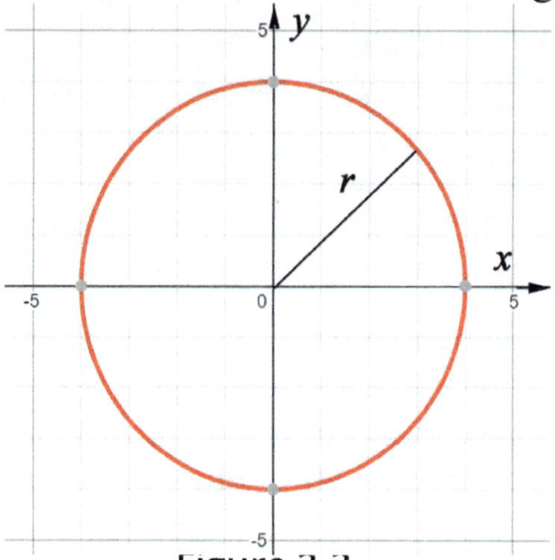

Figure 2-3

Consider the "*unit circle*," that is, a circle with radius 1:

$$x^2 + y^2 = 1^2 = 1$$

the "radius" will be stretched along the x-axis by a factor of a (the "radius" along the y-axis will remain 1). Figure 2-4 below shows the result for the case $a = 2$. The same effect will result along the y-axis when y is divided by b^2, where $b > 0$.

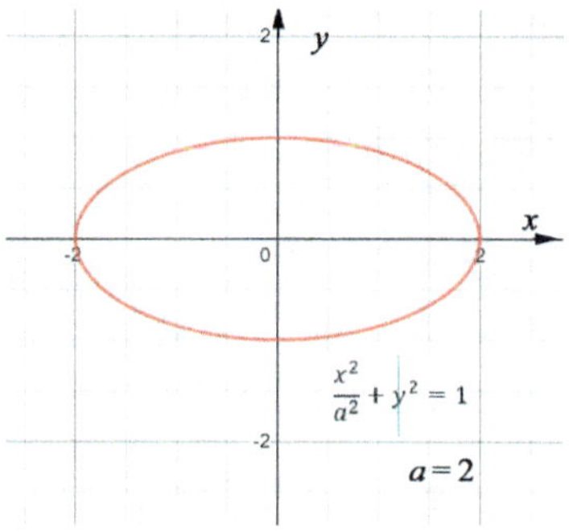

Figure 2-4

Taken together, the result is Equation (2-4), the equation of the ellipse.

$$\frac{x^2}{a^2} + \frac{y^2}{b^2} = 1 \qquad (2\text{-}4)$$

where a is the length of the *major axis* and b that of the *minor axis*.

Put Equation (2-4) into the form:

$$\frac{(x-h)^2}{a^2} + \frac{(y-k)^2}{b^2} = 1 \qquad (2\text{-}5)$$

Equation (2-5) is the equation of an *ellipse* centered at (h, k) with major axis a and minor axis b.

The basic conics equation is

$$Ax^2 + Bxy + Cy^2 + Dx + Ey + F = 0,$$

where A, B, and C are not all equal to zero. The "B" term involves rotation of the axes and is not considered here, so we set B = 0.

Ellipse:

$$Ax^2 + Cy^2 + Dx + Ey + F = 0$$

$$AC > 0 \text{ (A and C have like signs)}$$

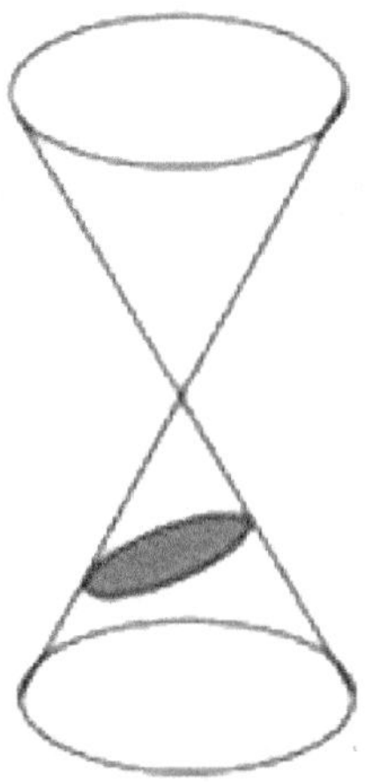

Ellipse

In my high school mechanical drawing class (1946) we learned to draw the ellipse using the "pin-and-string" method (this was in the 1940s – long before graphic calculators and digital computers): draw a line to establish the axis then push two pins into the drawing board. Take a piece of string and form a loop large enough to go loosely around the two pins. Put the pencil tip against the string and stretch it taut, then draw the ellipse by keeping the string taut. If you're steady you will reach the original point and have drawn an ellipse with its foci at the two pin position. You'll see why this is so as we discuss the conic section formed by a plane passing through a cone at an angle; of course, if the angle is zero, you'll have a circle which is really just a special case of the ellipse.

Terminology relating to the ellipse
(refer to Figure 2-5 below):

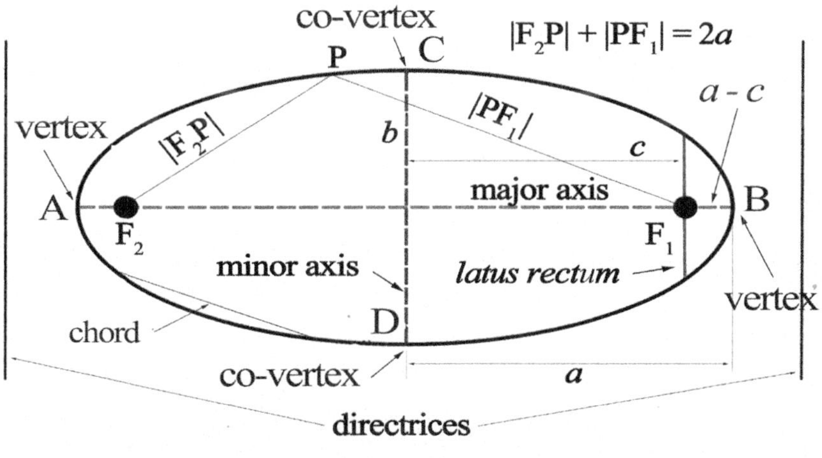

Figure 2-5

chord – a line segment having its endpoints on the ellipse.

co-vertex – an endpoint of the minor axis of the ellipse.

directrices - two lines parallel to the latus rectum and perpendicular to the major axis of the ellipse. The directrices are equidistant from the minor axis.

eccentricity - expressed as $e = c/a$ ("e" here is not Euler's constant). Figure 2-5 shows the distance, c, of a focus from the center is clearly less than a, that of a vertex; therefore $c < a$, and $e < 1$. If an ellipse's foci are moved toward the center, the ellipse approaches the shape of a circle, and if $c = 0$, then $e = 0$; that is, the foci merge into a center and the shape becomes a circle. Therefore, $0 < e < 1$ for any ellipse, and a circle is really a special type of ellipse.

focus (plural ***foci***) - two points on the major axis whose sum of distances from any point on the ellipse is always the same. The distance, denoted by c, between each focus and the center is the focal length of the ellipse.

latus rectum - a line passing through the foci of a conic that is parallel to the directrix of that conic. The *latus rectum* is the focal chord and the number of *latus rectums* is equal to the number of foci in the conic. The ellipse has two *latus rectums*. The length of the ellipse's *latus rectum* is $(2b^2)/a$.

line segments - The lengths of line segments *a, b, c* are measured from the center of the ellipse:

a – to either vertex.
b – to either co-vertex.
c – to either focus.

major axis – the axis passing through the foci and center and vertices of the ellipse.

minor axis – the axis passing through the center and co-vertices of the ellipse.

vertex – an endpoint of the major axis of the ellipse.

An ellipse is the set of all points in a plane such that the sum of the distances from two fixed points (foci) is constant. As shown in Figure 2-5, let the distance from each focus F_1 and F_2 to a given point P. The sum of the distances $| F_1P |$ and $| F_2P |$ from point P to the two foci is equal to $2a$, the length of the major axis between vertices A and B.

The conics form of the equation of the ellipse is where a is the length of the *major axis* and b that of the *minor axis*. How about length c, the distance between the foci and the center of the ellipse? Let's look at the lengths of line segments *a, b*, and *c*, referring to Figure 2-5. The minor axis, between C and D has length $2b$. Note that the focal length (between the focus and nearest vertex) has length $a - c$. The distance between vertices: $|AB| = c + a + (a - c) = 2a$

Definition of ellipse: $|F_1C| + |F_2C| = 2a$ and $|F_1C| = |F_2C| = |FC|$

EXAMPLES

Example 1. Give the eccentricity, lengths of both axes and coordinates of the vertices, co-vertices, foci, and center of the following ellipse:

$$9(x-2)^2 + 25(y+4)^2 = 225 \qquad \text{(E-1)}$$

• • Let's put the above into conics form of the ellipse

$$\frac{(x-\mathrm{h})^2}{a^2} + \frac{(y-\mathrm{k})^2}{b^2} = 1$$

by dividing equation E-1 through by 225 and then reducing to lowest terms:

$$\frac{9(x-2)^2}{225} + \frac{25(y+4)^2}{225} = 1$$

$$\frac{(x-2)^2}{25} + \frac{(y+4)^2}{9} = 1$$

The "standard" equation for the ellipse is

$$\frac{(x-2)^2}{5^2} + \frac{(y+4)^2}{3^2} = 1$$

Compare with

$$\frac{(x-\mathrm{h})^2}{a^2} + \frac{(y-\mathrm{k})^2}{b^2} = 1$$

and we see that h = 2, k = -4, $a = 5$, $b = 3$, and since $c^2 = a^2 - b^2 = 25 - 9 = 16$, we have c = 4 and its eccentricity is $\boldsymbol{e}$ = c/a = ⁴⁄₅. The length of the major axis is $2a = 10$ and that of the minor axis is $2b = 6$. We still need to determine the coordinates. The center of this ellipse is (2, -4). The foci are at distance c = 4 from the center so their coordinates are (2 – 4, -4) and (2 + 4, -4) or (-2, -4) and (6, -4). The vertices are at distance a from the center: (-3, -4) and (7, -4). The co-vertices are at distance b = 3 from the center at (2, -1) and (2, -7).

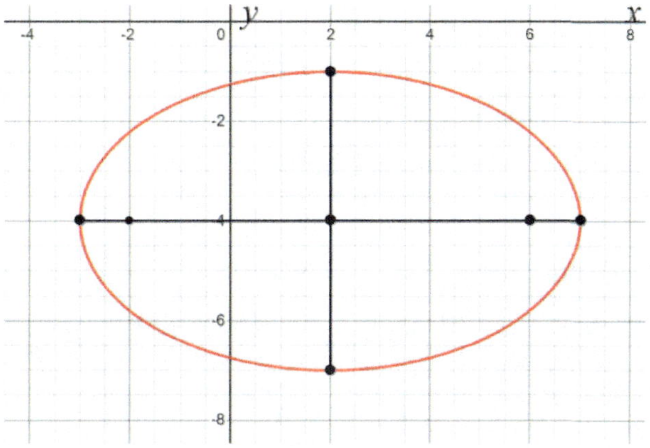

Example 2: An ellipse has a vertex at (2, 5), a focus at (2, 4), and is centered at (2, -2). Give the equation for this ellipse.

• • The given focus and vertex are lined vertically; i.e., have the same value of *x*. The other focus has y-coordinate six units below the center, or at $y = -8$, so its coordinates are (2, -8). The lower vertex is seven units away at coordinates (2, -9). Therefore, $a = -2 - (-9) = 7$ and $b = -2 - (-8) = 6$. The equation is

$$\frac{(y+2)^2}{7^2} + \frac{(x-2)^2}{6^2} = 1 \; or \; \frac{(y+2)^2}{49} + \frac{(x-2)^2}{36} = 1$$

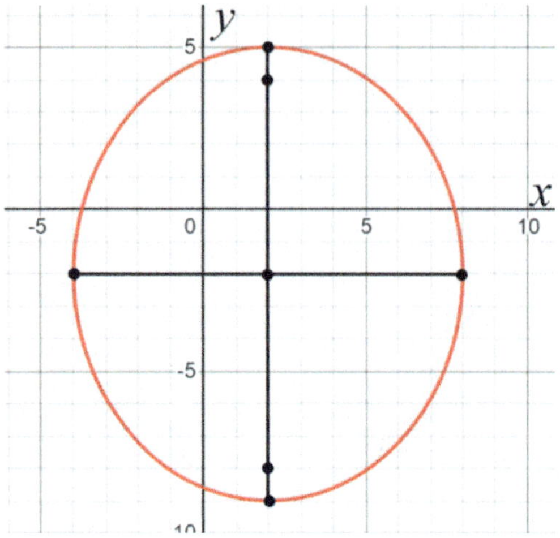

Example 3:

Give the equation of the ellipse centered at (2, 0) with vertex at (11, 0) and passing through the point (2, -5). What is the eccentricity of this ellipse?

• • We have $a = 11 - 2 = 9$. Write the equation with the given data:

$$\frac{(2-2)^2}{9^2} + \frac{(-5)^2}{b^2} = 1$$

from which *b* is easily calculated: $b = 5$.

$$\frac{(x-2)^2}{81} + \frac{y^2}{25} = 1$$

and $c^2 = a^2 + b^2 = 81 + 25 = 106$. The eccentricity is

$$\boldsymbol{e} = c/a = 10.3/9 = 1.1$$

approximately.

Example 4: An ellipse has foci at (0, -3) and (0, 3) and eccentricity $e = ½$.

• • Clearly, $c = 3$, the distance of each focus from the origin.

Because $\boldsymbol{e} = c/a$, we have

$$1/2 = 3/a \text{ from which } a = 6$$

The co-vertex distance from the origin is *b*, so

$$b^2 = a^2 - c^2 = 36 - 9 = 27.$$

There is enough data to write the equation of this ellipse:

$$\frac{x^2}{36} + \frac{y^2}{27} = 1$$

Parabola

Parabola

Basic equation:

$$Ax^2 + Bxy + Cy^2 + Dx + Ey + F = 0,$$

where A, B, and C are not all equal to zero. The "B" term involves rotation of the axes and is not considered here, so we set B = 0

Parabola: $Ax^2 + Cy^2 + Dx + Ey + F = 0$

where AC = 0 (A = 0 or C = 0 but not both)

The parabola is a conic section formed by a plane intersecting a cone at an angle parallel to the side of the cone. It is "U" shaped with a clear vertex (the only extreme point of the curve). A parabola is defined as the set (locus) of points that are equidistant from a point called the focus and a line perpendicular to the axis of symmetry of the parabola. The familiar quadratic equation

$$ax^2 + bx + c = 0$$

is that of a parabola, and is one of several equation forms for that curve. Let's derive another equation from this definition for a parabola.

Let the coordinates of the focus of the parabola be (0, p) = (0, 2). From the definition, the equation of the directrix is $y = -2$, as shown in Figure 2-6.

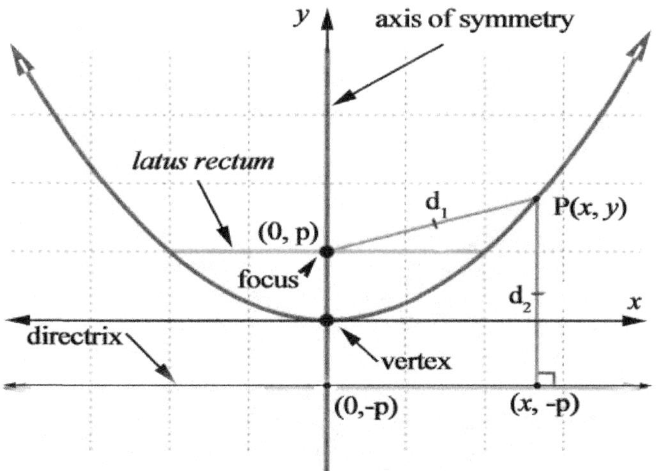

Figure 2-6

The definition states that the distance (call it d_1) from the focus to a point (x, y) on the parabola is equal to the perpendicular distance, (d_2), from (x, y) to the directrix (*i,e,,* the line $y = -2$). We use the familiar distance formula:

$$d_1 = \sqrt{(y-2)^2 + (x-0)^2} = \sqrt{(y-2)^2 + x^2}$$

and $$d_2 = \sqrt{[y-(-2)]^2 + (x-x)^2} = \sqrt{(y+2)^2}$$

Since $d_1 = d_2$, we equate the right-hand sides and square both sides to get

$$(y-2)^2 + x^2 = (y+2)^2$$

Expanding

$$y^2 - 4y + 4 + x^2 = y^2 + 4y + 4$$

Cancelling and collecting terms yields $x^2 = 8y$ or $y = x^2/8$ and since $p = 2$, we have an equation of the parabola

$$y = \frac{x^2}{2p} \text{ or } y = \frac{(x-h)^2}{2p} + k \text{ for vertex at } (h, k).$$

Terminology relating to the parabola
(see Figure 2-7)

axis of symmetry – the line passing through the focus and vertex of the parabola.

chord – a line segment whose endpoints lie on the parabola.

directrix – a line exterior to the parabola perpendicular to the axis of symmetry and a distance away from the vertex equal to the focal distance (the distance of the focus from the vertex).

focal chord – a chord that passes through the focus of the parabola.

focal distance – the distance from the focus to a point on the parabola; this is also the perpendicular distance from that point to the directrix.

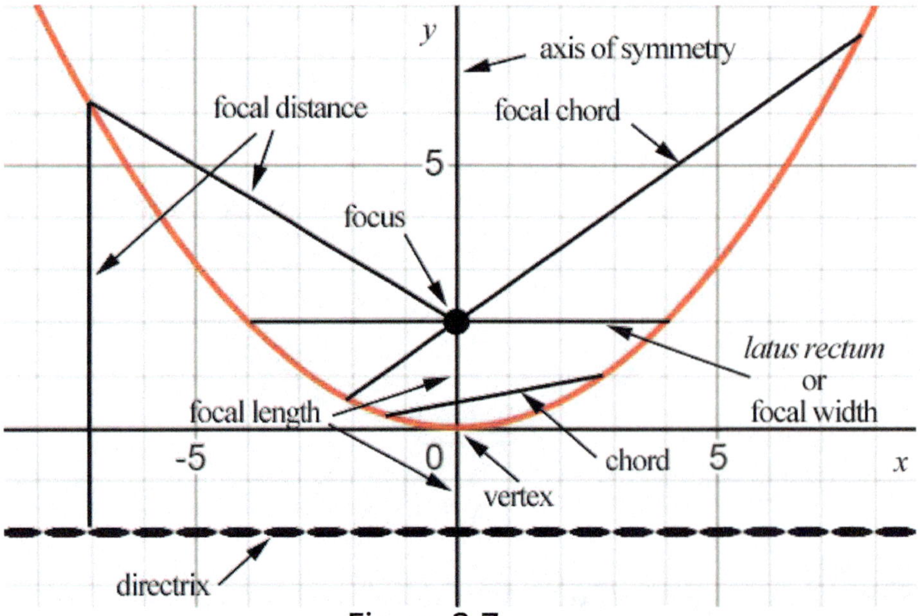

Figure 2-7

focal length – the focal distance to the vertex of the parabola.

focal width – the focal chord that is parallel to the directrix (also perpendicular to the axis of symmetry; also called the *latus rectum*.

focus – a fixed point on interior of a parabola located on the axis of symmetry such that the distance from the focus to any point on the parabola (called the "focal distance") is equal to the perpendicular distance of that point to the directrix.

latus rectum – focal width; see the definition above.

vertex – the *extremum* of the parabola that intersects the axis of symmetry.

Figure 2-6 shows the y-axis as its axis of symmetry and so the directrix is parallel to the x-axis. The equation $y^2 = 4px$ is a parabola symmetric about the x-axis and directrix parallel to the y-axis. In both cases the constant p is the distance from the vertex to the focus (also the distance from the vertex to the directrix).

Consider a parabola symmetric about the y-axis with vertex at (0, 0); i.e., $y = ax^2$ where a is the "steepness" of the arms of the parabola: $0 < a < 1$ is "flatter" and $x > 1$ "steeper" than x^2. As with any other function f(x), the expression f(x – h) + k is the function f(x) translated horizontally (parallel to the x-axis) by h units and vertically (parallel to the y-axis) by k units. For the parabola we have the so-called vertex form of the equation:

$$y = a(x - h)^2 + k$$

Example 1. Find the vertex and the coordinates of the focus of the parabola with equation $y + 8 = (x - 2)^2/12$. Give the equation of the directrix.

• • This is the conics form of the parabola. The function is translated 2 units in the x-direction and -8 units in the y-direction, giving the coordinates of the vertex as (2, -8).* The denominator of the right-hand side of the equation is 12. Calculate $2p = 12$ from which the focus, $p = 6$, is at coordinates (2, -2). The directrix is the line y = -14.

Example 2: Find the coordinates of the vertex and focus and equation of the directrix of

$$y^2 + 4y + 4x - 16 = 0$$

• • The squared term is y so this is a parabola with axis of symmetry parallel to the x-axis. Put the x term on the right side of the equation

$$y^2 + 4y - 16 = -4x$$

Now, complete the square:

$$y^2 + 4y + 4 - 16 = -4x + 4$$
$$(y + 2)^2 - 16 = -4x + 4$$
$$(y + 2)^2 = -4x + 20 = -4(x - 5)$$
$$x = (y + 2)^2/(-4) + 5 \quad \text{[vertex form]}$$

* The vertex of the parent function $y = x^2$ is the origin, (0, 0).

Compare the intermediate result

$$(y + 2)^2 = -4x + 20 = -4(x - 5)$$

with the conics form of the parabola having vertex at (h, k)

$$(y - k)^2 = 4p(x - h)$$

and we see that

$$h = 5,\ k = -2, \text{ and } p = -1.$$

See Figure 2-8.

Because the coefficient of the x-term is negative the parabola opens to the left. The coordinates of the vertex are (5, -2); the focus is two units to the left of the vertex (focus is always inside the parabola) with coordinates (5 – 1, -2) or (4, -2). The directrix is to the right of the left-opening parabola and is $x = 5 + 1 = 6$. Note that this graph is not a function because it fails the "vertical line" test.

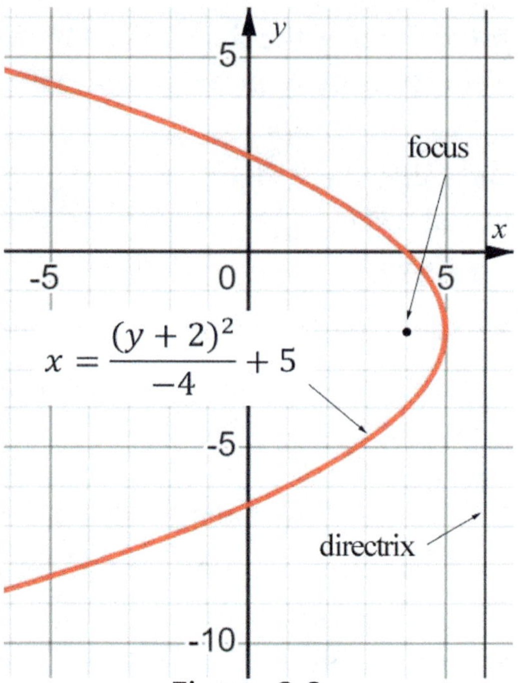

Figure 2-8

Hyperbola

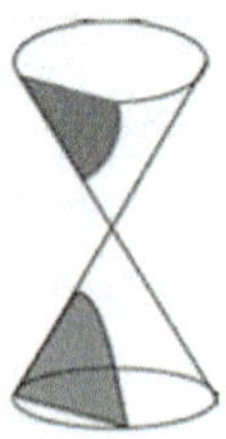

Hyerbola

Basic equation:

$$Ax^2 + Bxy + Cy^2 + Dx + Ey + F = 0,$$

where A, B, and C are not all equal to zero. The "B" term involves rotation of the axes and is not considered here, so we set B = 0

$$\textit{Hyperbola}:\quad Ax^2 + Cy^2 + Dx + Ey + F = 0 \qquad AC < 0$$

What a difference a *difference* makes! Change the " + " to a " – " in the conics form of the equation of an ellipse and and you have a hyperbola (if you don't pronounce the 'h' then it's "an hyperbola"). First, an example:

Example 1 on page 2-9 displays the equation of an ellipse, and now, the same terms of that equation separated by a minus sign give a hyperbola:

$$\frac{(x-2)^2}{5^2} + \frac{(y-4)^2}{3^2} \qquad \frac{(x-2)^2}{5^2} - \frac{(y-4)^2}{3^2}$$

ellipse hyperbola

You would think that these two expressions are related, so let's plot them on the same axis as shown in Figure 2-9:

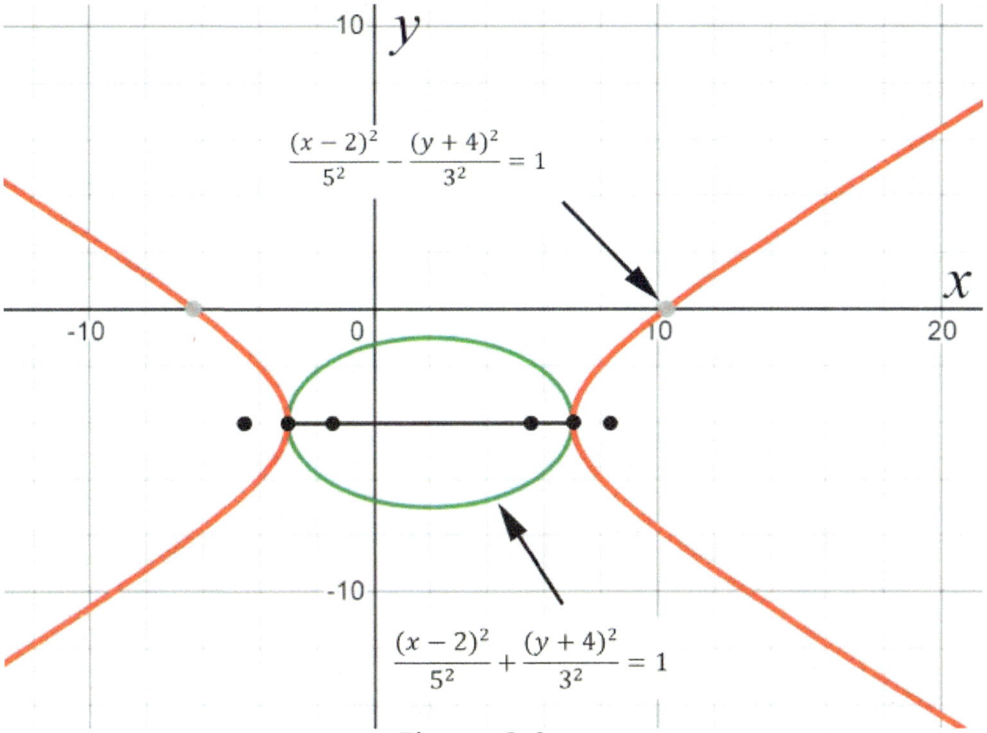

Figure 2-9

Recall that for an ellipse the sum of the distances from each focus to a point on the ellipse is constant, and derived the equation

$$\frac{x^2}{a^2} + \frac{y^2}{b^2} = 1$$

If the difference of distances from each focus to a point is a constant then we have a hyperbola.

$$\frac{x^2}{a^2} - \frac{y^2}{b^2} = 1$$

Both the ellipse and hyperbola have two foci and two vertices (only the ellipse has co-vertices). A diagram showing the nomenclature of the hyperbola is shown in Figure 2-10. Note that the transverse axis length is $2a$ and that of the conjugate axis is $2b$. The slopes of the diagonals of the "fundamental" or "central" box formed by these axes are b/a and $-b/a$, respectively. Note also that the distance of each vertex from the center is "a" and that of each focus is "c," the length of the diagonal of the "box." Therefore, $c^2 = a^2 + b^2$. Extending the diagonals forms lines that are the four asymptotes of the parabola.

Terminology relating to the hyperbola (refer to Figure 2-10):
axes of symmetry – there are two: the lines passing through the transverse axis and conjugate axis, respectively.

center - the midpoint of the transverse axis, where it is intersected by the conjugate axis

conjugate axis – the line segment of length 2*a* that bisects the transverse axis.

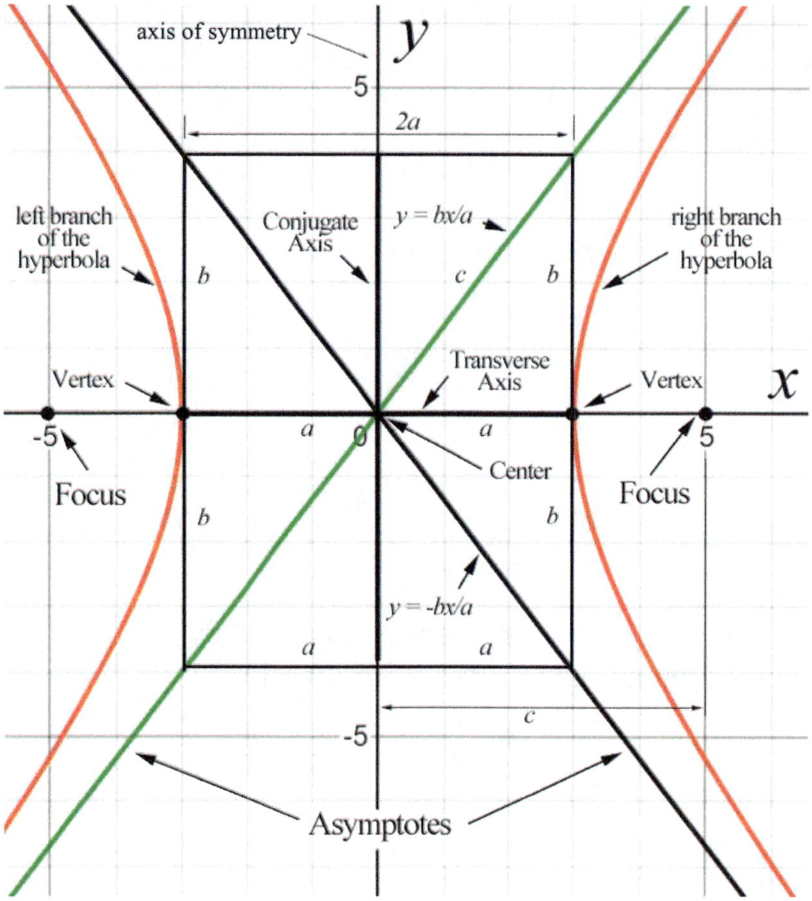

Figure 2-10

focus – one of two foci, call them F_1 and F_2, of the hyperbola. The difference of the distance from one of the foci to any point on the hyperbola and the distance from that point to the other focus is constant.

fundamental box (see Figure 2-10) - also called the "central box," the rectangle centered on the hyperbola origin. Its length (parallel to the transverse axis) is 2*a* and height (parallel to the conjugate axis) is 2b.

line segments a, b, c -

a - the distance from the origin to either vertex of the hyperbola.

b - the length of the conjugate axis

c - determined by the Pythagorean relationship: $c^2 = a^2 + b^2$.

right/left branch of the hyperbola (also called "arm of the hyperbola") - the portion of the parabola that is to the right/left of the conjugate axis.

transverse axis – the line segment passing through the origin and having end-points on the vertices of the hyperbola.

vertex **-** the point where the axis of the hyperbola intersects the hyperbola.

More terminology relating to the hyperbola
(shown in Figures 2-11 and 2-12 below)

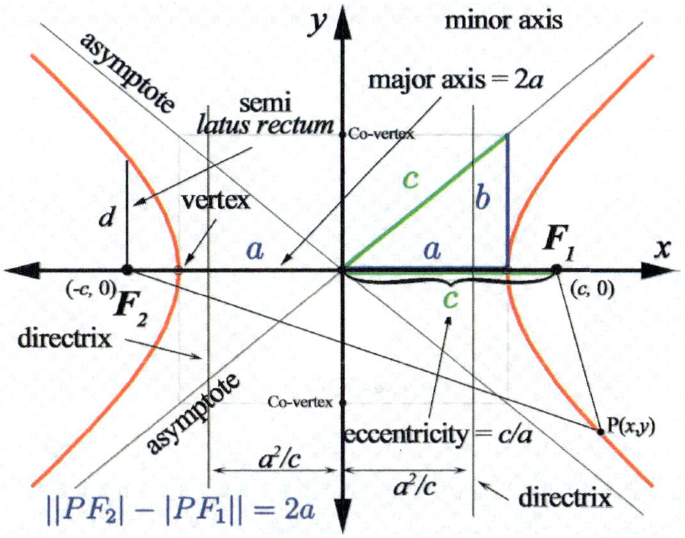

Figure 2-11

directrix **-** the line which, together with the point known as the focus, serves to define a conic section as the locus of points whose distance from the focus is proportional to the horizontal distance from the directrix, with $r > 1$, the constant of proportionality. The hyperbola has two distinct foci and two associated directrices that are perpendicular to the line joining the two foci.

chord – A straight line segment whose endpoints both lie on a circular arc, such as that of a circle, parabola, or hyperbola. For both the parabola and hyperbola the line segment lies in the concave part of the graph. See Figure 2-12 on the next page.

eccentricity - Eccentricity of the hyperbola: This is given by $e = c/a$. As the foci distance becomes larger, the hyperbola branches will "open."

focal chord – a chord that passes through a focus of the hyperbola.

focal distance – the distance of any line segment with one endpoint on a focus and the other its arm of the hyperbola.

major axis – the line passing through the foci, origin, and vertices of the hyperbola. The transverse axis lies on the major axis.

focal width - Also *latus rectum*. In a conic section the focal width is a chord parallel to the directrix that passes through a focus of the hyperbola.

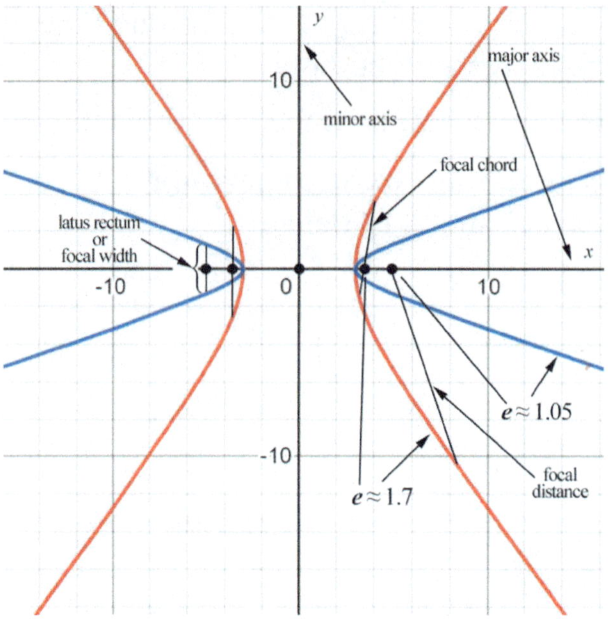

Figure 2-12

latus rectum (also "focal width") - a line of length $b^2/2a$ passing through the foci of the conic parallel to the directrix of the conic. It is a focal chord and the number of *latus rectums* is equal to the number of foci in the conic.

minor axis – the line passing through the origin and perpendicular to the major axis of the hyperbola.

Conjugate hyperbola

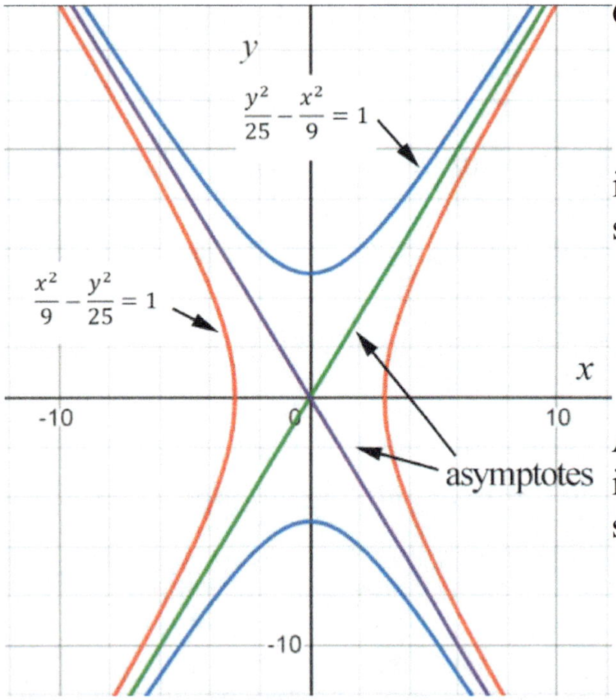

Figure 2-13

Given a hyperbola with equation

$$\frac{x^2}{a^2} - \frac{y^2}{b^2} = 1$$

interchange the two terms on the left side of the equation to get the

$$\frac{y^2}{b^2} - \frac{x^2}{a^2} = 1.$$

equation of the *conjugate hyperbola*: A hyperbola and its conjugate have identical asymptotes. The hyperbola shown in Figure 2-13 is

$$\frac{x^2}{9} - \frac{y^2}{25} = 1;$$

its conjugate $\frac{y^2}{25} - \frac{x^2}{9} = 1.$

Rectangular hyperbola

If a = b then the hyperbola is said to be rectangular or equilateral. Its asymptotes are perpendicular to each other. See Figure 2-14. Try graphing the function $y = 1/x$, which is a rectangular hyperbola having the x and y axes as its asymptotes (note that the main axis is the line $y = x$ and the minor axis is $y = -x$).

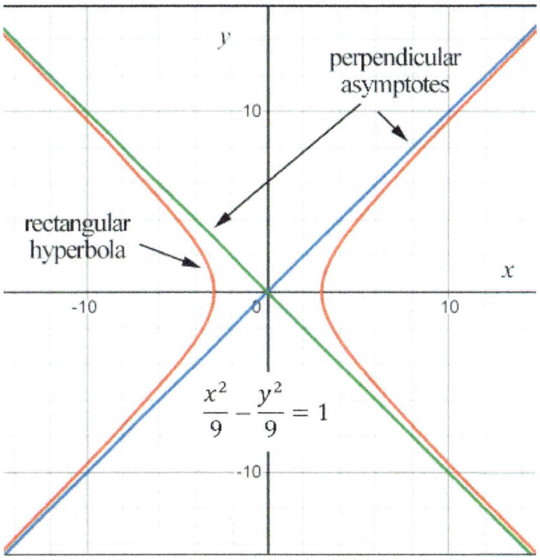

Figure 2-14

Example 1: Find the center, vertices, foci, eccentricity, asymptotes and conjugate hyperbola for the hyperbola

$$\frac{x^2}{5^2} - \frac{y^2}{2^2} = 1$$

• See Figure 2-15 below. The first term is the x-term so the parabola opens to the right and left; the center, vertex and foci are on the x-axis, the major axis. The center is (0, 0), the vertices are at (-5, 0) and (5, 0); the foci are at (-√41, 0) and (√41, 0). From the equation above we see that a = 5 and b = 2 so $c^2 = 25 + 4 = 29$ and the eccentricity is

$$e = \frac{\sqrt{29}}{5} \approx 1.08$$

Finally, the asymptotes are $y = 2x/5$ and $y = -2x/5$.

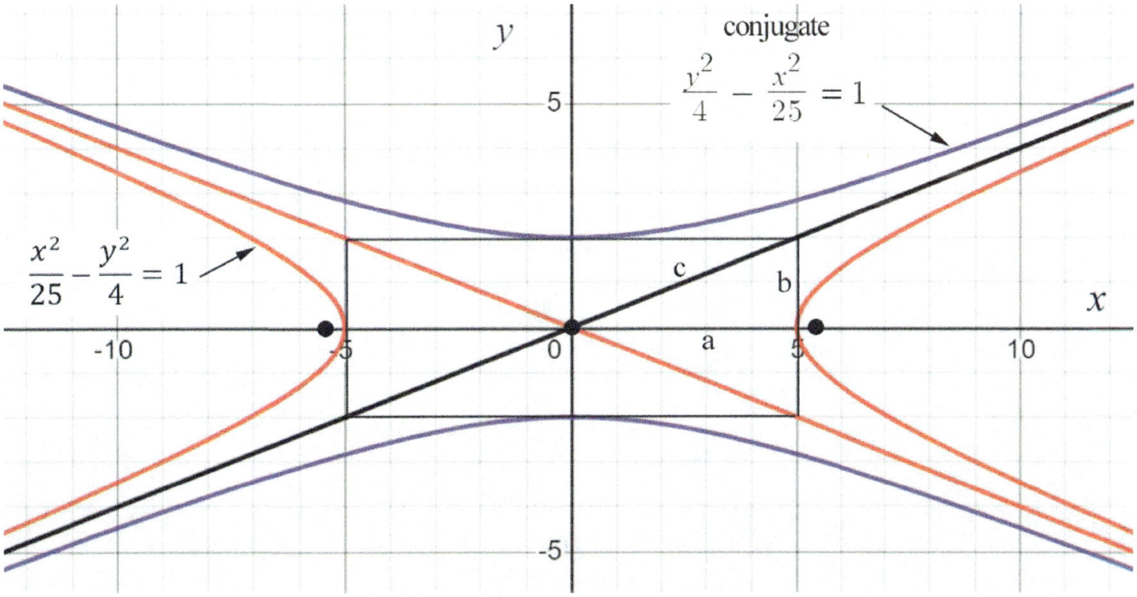

Figure 2-15

Example 2. Find the center, vertices, foci, and asymptotes of the hyperbola

$$3x^2 - 3y^2 + 6x + 6y - 75 = 0.$$

• Simplify first to get

$$x^2 + 2x - (y^2 - 2y) - 25 = 0.$$

To get the above equation in conics form we next complete the square:

$$(x^2 + 2x + 1) - (y^2 - 2y + 1) - 25 = 0 + 1 - 1 = 0$$

$$(x + 1)^2 - (y - 1)^2 = 25$$

$$\frac{(x+1)^2}{25} - \frac{(y-1)^2}{25} = 1$$

We see immediately that this is a rectangular hyperbola centered at (-1, 1); $a = b = 5$; and $c = \sqrt{50}$. The asymptotes are 1 and -1 and are perpendicular.

Example 3. Find the equation in the form $Ax^2 + Cy^2 = Dx + Ey + F = 0$ for the hyperbola centered at (3, 5), vertex (3, 5), and focus (11, 5).

• The major axis for this hyperbola is $y = 5$. The vertex is three units from the center so $a = 6$ and $c = 8$, the displacement from the center of the foci. Calculate $b^2 = 36 + 64 = 100$ so $b = 10$.

The conics form of this hyperbola is $\frac{(x-3)^2}{36} - \frac{(y-5)^2}{64} = 1$

To get this in the desired form, we must expand the squares after getting a common denominator. The least common denominator is 576 (see the computations below*) so the equation becomes

$$\frac{16(x-3)^2}{576} - \frac{9(y-5)^2}{576} = 1$$

Multiply by 576 then expand the squares

$$16(x^2 - 6x + 9) - 9(y^2 - 10y + 25) = 576$$

Clear the parentheses

$$16x^2 - 96x + 144 - 9y^2 + 90y - 225 = 576$$

Place in the standard format: $Ax^2 + Cy^2 = Dx + Ey + F = 0$

$$16x^2 - 9y^2 - 96x + 90y + (144 - 225 - 576) = 0$$

$$16x^2 - 9y^2 - 96x + 90y - 657 = 0$$

So A = 16; C = -9; D = -96; E = 90 and F = -657

* Factor 36 and 64 into their prime factors to get 2, 2, 3, 3 and 2, 2, 2, 2, 2, 2 respectively. Display the factors in a table aligning common factors in columns as shown. Then List all of the factors (common factors only once):

64	2	2	2	2	2	2			
36	2	2					3	3	
	2	2	2	2	2	2	3	3	= ***576*** (product of these factors is the LCD)

Multiply the numbers in the bottom row to get the least common denominator (greatest common multiple).

Summary of conic sections

Basic equation:

$$Ax^2 + Bxy + Cy^2 + Dx + Ey + F = 0,$$

where A, B, and C are not all equal to zero. The "B" term involves rotation of the axes and is not considered here, so set B = 0.

Parabola:	$Ax^2 + Cy^2 + Dx + Ey + F = 0$	$AC = 0$	(A = 0 or C = 0 but not both)
Ellipse:	$Ax^2 + Cy^2 + Dx + Ey + F = 0$	$AC > 0$	(A and C have like signs)
Circle:	$Ax^2 + Cy^2 + Dx + Ey + F = 0$	$A = C$	
Hyperbola:	$Ax^2 + Cy^2 + Dx + Ey + F = 0$	$AC < 0$	

●

Summary: parabola, ellipse, and hyperbola

PARABOLA		
$Ax^2 + Cy^2 + Dx + Ey + F = 0$ where $AC = 0$ (A = 0 or C = 0 but not both)		
Conic equation	$(x - h)^2 = 4p(y - k)$	$(y - k)^2 = 4p(x - h)$
Vertex	(h, k)	(h, k)
Focus	(h, k + p)	(h + p, k)
Directrix	$y = k - p$	$x = h - p$
Axis of symmetry	$x = h$	$y = k$
Focal length	p	p
Focal width	4p	4p
Eccentricity	$e = 1$	$e = 1$
Opens …	$p > 0$, up; $p < 0$; down	$p > 0$, right; $p < 0$, left

ELLIPSE		
$Ax^2 + Cy^2 + Dx + Ey + F = 0$ $AC > 0$ (A and C have like signs)		
Equation	$\frac{(x-h)^2}{a^2} + \frac{(y-k)^2}{b^2} = 1$	$\frac{(y-k)^2}{a^2} + \frac{(x-h)^2}{b^2} = 1$
Center	(h, k)	(h, k)
Vertices	$(h \pm a, k)$	$(h, k \pm a)$
Co-vertices	$(h, k \pm b)$	$(h \pm b, k)$
Focal axis	$y = k$	$x = h$
Foci ($c^2 = a^2 - b^2$)	$(h \pm c, k)$	$(h, k \pm c)$
Semi-major axis	a	a
Semi-minor axis	b	b
Eccentricity	$e = c/a$	$e = c/a$
Pythagorean relation	$a^2 = b^2 + c^2$	$a^2 = b^2 + c^2$

HYPERBOLA		
$Ax^2 + Cy^2 + Dx + Ey + F = 0$ $AC < 0$		
Conic equation	$\frac{(x-h)^2}{a^2} - \frac{(y-k)^2}{b^2}$	$\frac{(y-k)^2}{a^2} - \frac{(x-h)^2}{b^2}$
Coordinates of the center	(h, k)	(h, k)
Coordinates of the vertices	$(h \pm a, k)$	$(h, k \pm a)$
Coordinates of the foci	$(h \pm c, k)$ where	$(h, k \pm c)$ where
Length of transverse axis	$2a$	$2a$
Length of conjugate axis	2b	2b
Eccentricity	$e = c/a$	$e = c/a$
Latus rectum	$2b^2/a$	$2b^2/a$
Asymptotes	$y = \pm\frac{b}{a}(x - h) + k$	$y = \pm\frac{a}{b}(x - h) + k$

THE
MATH TUTOR'S FRIEND
$z = a + ib$
i
imaginary unit
It's easy !!
Re - axis
real number
Im - axis
Mr. Math
COMPLEX
NUMBERS

THE MATH TUTOR'S FRIEND
$z = a + ib$
i
imaginary unit
It's easy !!
Re - axis
real number
Im - axis
Mr. Math
COMPLEX NUMBERS

3. COMPLEX NUMBERS

TABLE OF CONTENTS

page

•

3. COMPLEX NUMBERS

This section deals with the *complex number* which consists of a *real part* and an *imaginary part*. Before proceeding let's ask the questions "What is a *real number*?" and "What is an *imaginary number*?"

Real numbers

Natural numbers, whole numbers, integers, rational numbers, irrational numbers, and transcendental numbers are real numbers. Informally, a real number is any number on the number line and may be negative, positive, or zero: -238, -π, 0, ***e***, √2, and 17 are examples of real numbers.

Imaginary numbers

The Cartesian plane is characterized by a horizontal *x*-coordinate and a vertical *y*-coordinate (see Figure 3-1). Now take this coordinate system and re-label the axes as follows: real numbers are measured along the horizontal direction (positive or negative) as distance from the vertical axis analogous to the *x*-value of the ordered pair (*x*, *y*) that represents a point. Call this the *Real axis*, denoted by ***ℜe*** instead of *x*. Imaginary numbers are measured in the vertical direction (positive or negative) as a distance from the horizontal axis (analogous to the *y*-value of a point (*x*, *y*). Label this imaginary axis as ***ℑm*** instead of *y*. See Figure 3-2. As with the Cartesian system, a point on the plane is defined by the ordered pair (*x*, ***i****y*); that is, the first member, *x*, of the ordered pair is the distance along the ***ℜe*** axis, and *y*, the second member, is the distance along the ***ℑm*** axis; this is indicated by preceding* *y* with ***i***. Call the plane with the real and imaginary axis the *Complex Plane*. Notice that both *x* and *y* are real numbers; the symbol ***i*** is the *imaginary unit*, or simply imaginary number that not only indicates the "*y*-direction" but is formally defined as ***i*** = √-1, or more understandably, $i^2 = -1$. Therefore, $(i3)^2 = (i^2 3^2) = -9$, and $(i3)^4 = (i^4 3^4) = i^4 \times 81$; but $i^2 = -1$ and $i^4 = (i^2)^2 = (-1)^2 = 1$, so, $(i3)^4 = 1 \times 81 = 81$.

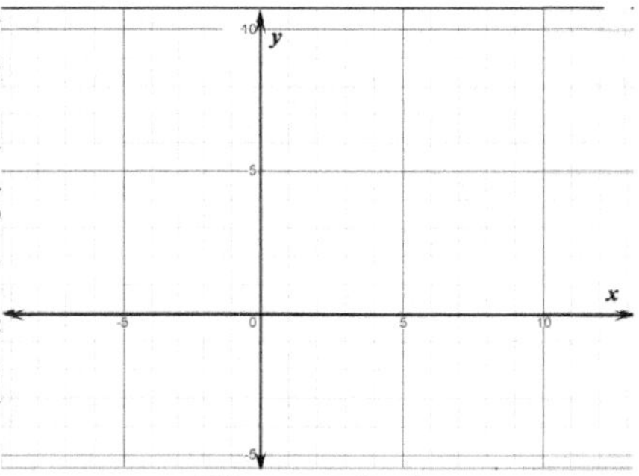

Figure 3-1

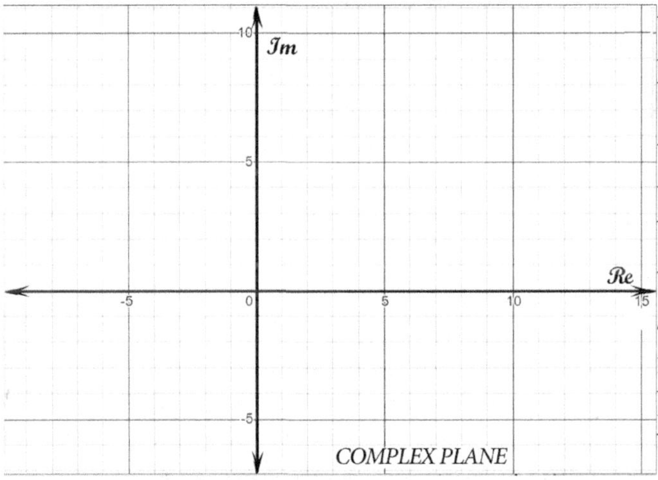

Figure 3-2

* Some texts and papers show the ***i*** following the number, such as "5***i***" instead of "***i***5" However, in trigonometry the Euler Formula is given as $e^{i\theta} = cos(\theta) + i\, sin(\theta)$ [= $cis(\theta)$] and so we use the same convention. In electrical engineering the imaginary unit ***j*** is used instead of ***i***, because ***i*** is the symbol used for electric current.

Complex number

Having described both real and imaginary numbers we now pull everything together:

$$z = a + \boldsymbol{i}b$$

where z is the *complex number* consisting of a *real part*, a, and an *imaginary part*, $\boldsymbol{i}b$, composed of the imaginary unit, $\boldsymbol{i}$, and a real number b. And, of course, $\boldsymbol{i}^2 = -1$. Refer to Figure 3-3.

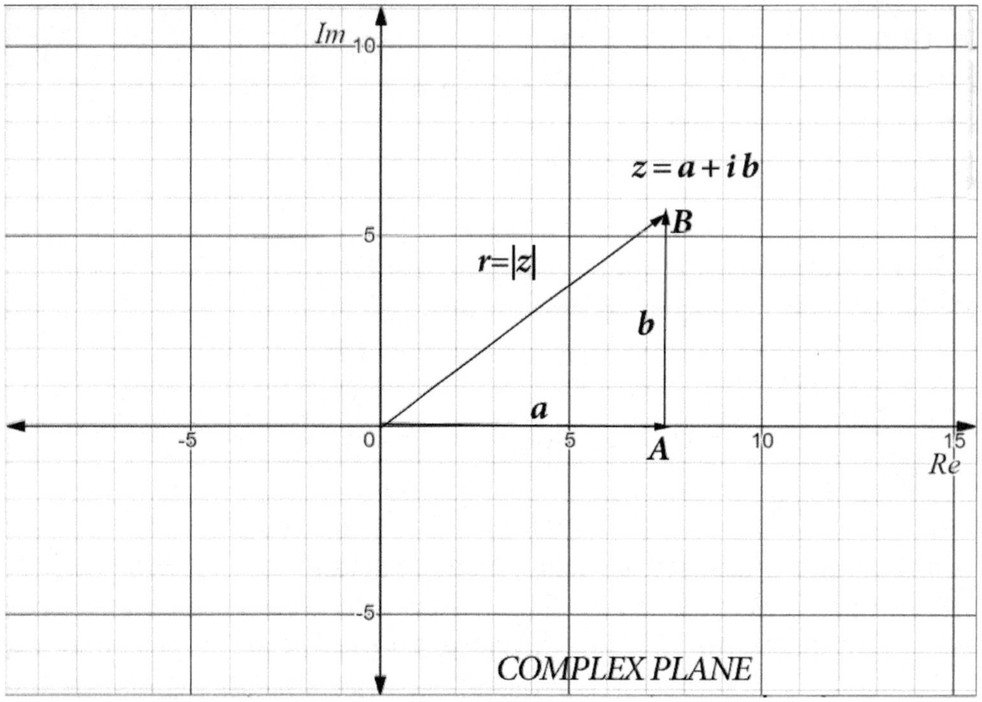

Figure 3-3

Every number can be expressed as a complex number. For example, 15 can be represented as $15 + \boldsymbol{i}0$, a complex number. The imaginary number $\boldsymbol{i}5 = 0 + \boldsymbol{i}5$, is a complex number also. Both Figures 3-3 and 3-4 show the point (a, $\boldsymbol{i}b$) on the complex plane.

Math operations with complex numbers

We'll work with the two complex numbers z and w, where

$$z = a + \boldsymbol{i}b \text{ and } w = c + \boldsymbol{i}d.$$

Absolute value of a complex number

The triangle ***AOB*** in Figure 3-3 is a right triangle with hypotenuse r and sides a and b, respectively - all real numbers. The absolute value, $|z|$, of a complex number is simply the distance of the point (on the complex plane) $z = a + \boldsymbol{i}b$ from the origin, ***O***. See Figure 3-4. The distance r may be calculated using either the *distance formula* or the *Pythagorean Theorem*: $|z| = \sqrt{(a^2 + b^2)}$.

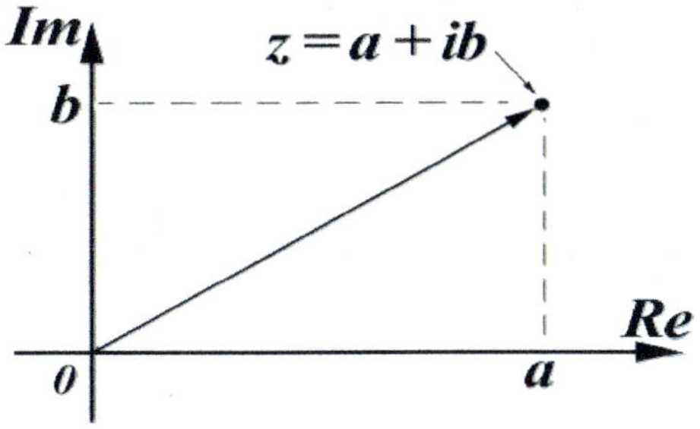

Figure 3-4

Adding complex numbers

Example: Calculate $z + w = (a + ib) + (c + id)$.

Simply add the real parts and add the real values of the imaginary parts:

$$z + w = (a + ib) + (c + id) = (a + c) + i(b + d)$$

and, of course,

$$z - w = (a + ib) - (c + id) = (a - c) + i(b - d).$$

Multiplying complex numbers

Example: Calculate $z \times w = (a + ib) \times (c + id)$.

The two terms in parentheses are binomials so we can use the FOIL method for multiplication:

$$z \times w = ac + iad + ibc + ib \times id \quad \text{[FOIL]}$$
$$= ac + i^2bd + i(ad + bc) \quad \text{[simplify]}$$
$$= ac - bd + i(ad + bc) \quad [i^2 = -1]$$

Dividing complex numbers

First, introduce the *complex conjugate* of $(a + ib)$. The complex conjugate of $z = (x + iy)$ is $\bar{z} = (x - iy)$; just use the opposite sign before the imaginary term.

See Figure 3-5.

Example: Divide $12 + i5$ by $3 - i$;
that is, $(12 + i5)/(3 - i) = ?$

Just as we do not want radicals in the denominator of a fraction, so too we do not want imaginary numbers in a denominator. The technique is to multiply the denominator and numerator by the conjugate of the denominator;

$$\frac{(12 + i5)}{(3 - i)} \times \frac{(3 + i)}{(3 + i)} = \frac{(12 + i5)(3 + i)}{(3 - i)(3 + i)}$$

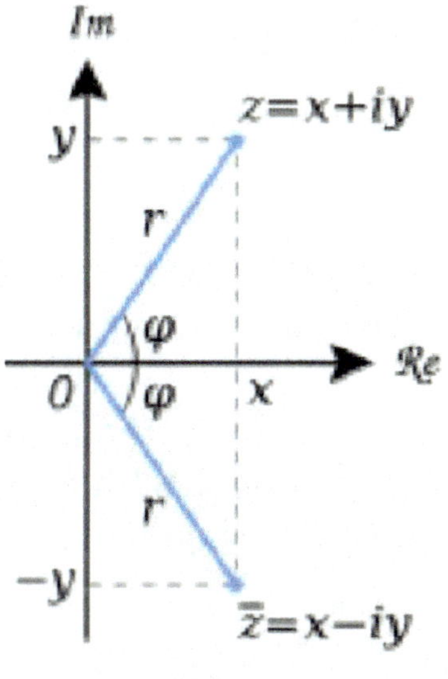

Figure 3-5

$(3 + i)$ is the conjugate of the denominator $(3 - i)$. Proceeding,
The pattern of the two conjugate binomials in the denominator, $(3 - i)(3 + i)$, is a “difference of squares” which results in

$$3^2 - i^2 = 9 - (-1) = 10.$$

Use FOIL for the numerator:

$$(12 + i5)(3 + i) = 12\times3 + 12i + 15i + i^2\times5 = 36 + 27i - 5 = 31 + 27i$$

all divided by 10. The result is $(31 + 27i)/10 = 3.1 + i2.7$.

Raising i to an integer power

Note that

$i^0 = 1$ [anything (but zero) to the zero power equals 1]

$i^1 = i$ [anything to the 1st power equals itself]

$i^2 = -1$ [definition of i]

$i^3 = i \times i^2 = i \times (-1) = -i$

and $i^4 = i^2 \times i^2 = (-1)\times(-1) = 1 = i^0$

See the pattern? All you need do is divide the exponent of i by 4 and look at the remainder, which will be 0, 1, 2, or 3, as shown above (this is “n mod 4”).

So,

$i^8 = i^0 = 1$ because $8/4 = 2$ with remainder 0

$i^{25} = i^1 = i$ because $25/4 = 6$ with remainder 1

$i^{26} = i^0 = -1$ because $26/4 = 4$ with remainder 2

$i^{75} = i^3 = -i$ because $75/4 = 18$ with remainder 3

Raising a complex number to a complex power

It is interesting to note that the imaginary unit “i” raised to the ith power results in a real number:

Take Euler’s formula (see Section 6);

$$e^{i\pi} + 1 = 0$$

add “-1” to both sides; take the square root of both sides (yielding i), then raise both sides to the power i. The sequence of steps is shown to the right; note that the result is the real number

$$e^{-\pi/2}.$$

$$e^{i\pi} + 1 = 0$$

$$e^{i\pi} = -1$$

$$e^{i\pi/2} = \sqrt{-1} = i$$

$$i^i = e^{\frac{i^2\pi}{2}} = e^{-\pi/2}$$

Complex numbers in equations

Consider the equation

$$x + iy = 25 + i(n + 4)$$

The solution is found by equating the real part on one side of the equation to the real part on the other side and to equate the coefficients of i on both sides of the equation. Therefore, the solution to the example is $x = 25$ and $y = n + 4$.

Complex factors of polynomials

Consider the two functions f(x) and g(x) shown below. Factor g(x) completely and then use synthetic division to show that one of the factors of g(x) is also a factor of f(x).

$$f(x) = x^5 - 4x^4 + (4 + 2i)x^3 - (16 + 8i)x^2 + 8ix - 32i$$
$$g(x) = x^2 + 2i$$

First, reformat g(x) into a difference of squares by noting that the term " $-2i$ " can be the middle term of a perfect-square trinomial:

$g(x) = x^2 + 2i$
$= x^2 + 2i = x^2 - (-2i)$ [replace "$+ 2i$" with "$-(-2i)$"]
$= x^2 - (1 - 1 - 2i)$ [add 1 and its opposite to $-2i$]
$= x^2 - (-1 - 2i + 1)$ [rearrange expression in parentheses]
$= x^2 - (i^2 - 2i + 1)$ [substitute i^2 for -1; this is a perfect square]
$= x^2 - (i - 1)^2$ [difference of squares pattern]
$= [x + (i - 1)]\times[x - (i - 1)]$ [expand into conjugate factors]

g(x) is now completely factored into a conjugate pair of complex expressions. Now, proceed to determine if one of these factors is also a factor of f(x). Do this using synthetic division of f(x) by $[x - (i - 1)]$.

	[Col 1	2	3	4	5	6]
	1	-4	$4 + 2i$	$-16 - 8i$	$8i$	$-32i$
$i - 1$		$i - 1$	$4 - 6i$	$-4 + 12i$	$16 - 24i$	$32i$
	1	$i - 5$	$8 - 4i$	$-20 + 4i$	$16 - 16i$	0

Procedure (lots of practice using F.O.I.L.):

1st column: copy the "1" and put the product of $(i - 1)\times 1$ in the
2nd column: add; put the product of $(i - 1)\times(i - 5) = (4 - 6i)$ in the
3rd column: add; put the product of $(i - 1)\times(8 - 4i) = (-4 + 12i)$ in the
4th column: add; put the product of $(i - 1)\times(-20 + 4i) = (16 - 24i)$ in the
5th column: add; put the product of $(i - 1)\times(16 - 16i) = 32i$ in the
6th column; adds to zero confirming that $(i - 1)$ is a factor of f(x).

•

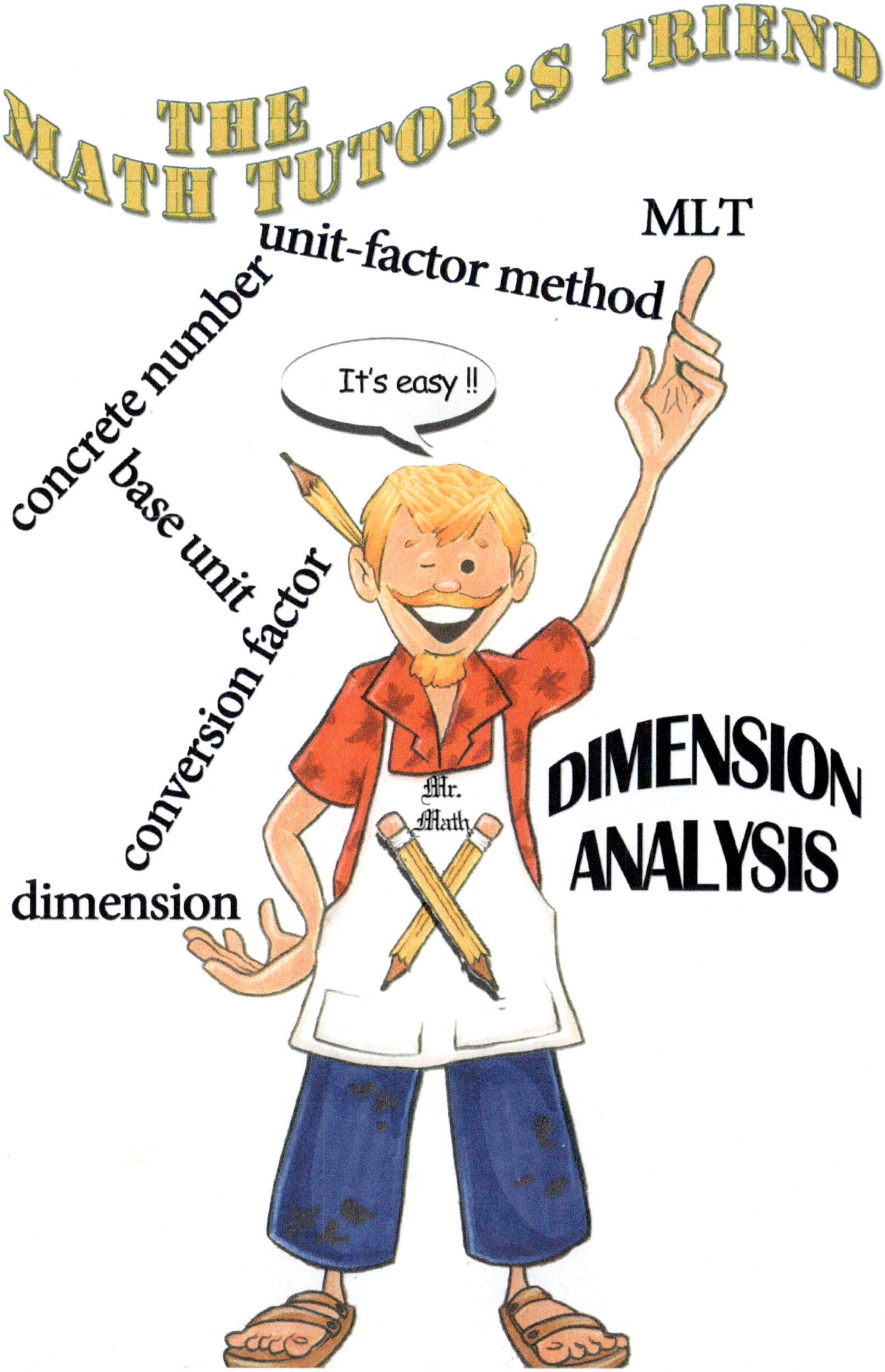
THE MATH TUTOR'S FRIEND
MLT
unit-factor method
concrete number
It's easy !!
base unit
conversion factor
dimension
Mr. Math
DIMENSION ANALYSIS

4. DIMENSION ANALYSIS
TABLE OF CONTENTS

page

•

4. DIMENSION ANALYSIS

Dimension analysis

Dimension analysis (also *dimension cancelling, dimensional analysis, unit-factor method, factor-label method*) examines the relationship among physical quantities by identifying their *base quantities* (length, mass, time, electric current*) and units of measure such as miles *vs* kilometers and pounds *vs* kilograms. We shall convert from one dimensional unit to another using algebraic techniques.

Physical quantities

Physical quantities such as feet and miles are measures of *length* (L); pounds and kilograms are measures of *mass* (M); and seconds and years are measures of *time* (T). Each pair cited above consists of *commensurable quantities* and can be compared to each other, for example, 1 hour = 3600 seconds. Divide both sides of this equality by 1 hour to get 1 hour/1 hour = 3600 seconds/1 hour. But 1 hour/1 hour = 1, so 3600 seconds/1 hour = 1 and we say that there are 3600 seconds in 1 hour. Thus, we can compare seconds, hours, minutes, days, years, centuries, …) as these quantities are commensurable. This is the basis for dimension analysis. We cannot compare 1 kilogram to 1 kilometer because they are *incommensurable* quantities. Therefore any equation or inequality must have commensurable quantities on both sides of the " = " " < " " > " "≥" and " ≤ " signs.

Checking computational results

Answers to algebraic problems must have commensurable quantities. That is, both sides of an equation must have the same dimensions (*dimensional homogeneity*). You might have an incorrect answer with commensurable equations but if your results equate incommensurable quantities you most certainly do have a wrong answer!

Commensurable quantities and mathematical operators

Just as you cannot add apples and oranges you cannot add incommensurable quantities. However, it is perfectly OK to multiply or divide incommensurable quantities (e.g., 90 miles/2 hours). In engineering and physics or any other quantitative science one deals with "constants" that may be dimensionless (example, π, %) or have dimension (example, G, the gravitational constant has dimensions

$$\text{meters}^3/(\text{kg} \times \text{seconds}^2) \text{ or } L^3/MT^2.$$

Base unit

A *base unit* is best described by an example; *length* (L) is a base unit which cannot be expressed in terms of other units. *Volume* (L^3) is a derived unit which can be factored into base units (i.e., L×L×L). Similarly, *mass* (M) and *time* (T) are base units.

* Electrical units are not addressed in this book.

Concrete number

A *concrete number* (in contrast with an *abstract number*) is associated with an object; for example, 3 feet ("3" by itself is an abstract number).

Compound relation

An example of a *compound relation* is velocity which has dimensions L/T, e.g., miles per hour ("per" implies division).

Conversion factor

The *conversion factor* is a ratio that converts one unit of measure to another without changing the quantity, for example 2π radians/360° = 1.

Dimensional homogeneity

This is a basic rule that states that only commensurable units may be compared, equated, added or subtracted. However, as we have seen, ratios or products of incommensurable units are allowed.

Factor-label method

The factor-label method is the crux of dimension analysis. It is a sequential application of conversion factors expressed as fractions. This is an application of dimension canceling, or dimension analysis used by scientists and engineers. It is based on the fact that if you have the same thing in both the numerator and denominator, the result is unity: 3/3 = 1; 100,000,000/100,000,000 = 1; $xy/xy = 1$; inches/inches = 1; liters/liters = 1. Got it?

A trillion seconds - how many years?

Example: How many years equals one trillion seconds? We wish to convert one trillion seconds to years. Let's pursue this example using the factor-label method. We'll use scientific notation to represent a trillion seconds: 1,000,000,000,000 = 10^{12}. Now to the problem:

$$\frac{10^{12}\ \textit{seconds} \times 1\ \textit{minute} \times 1\ \textit{hour} \times 1\ \textit{day} \times 1\ \textit{year} \times 1\ \textit{century}}{60\ \textit{seconds} \times 60\ \textit{minutes} \times 24\ \textit{hours} \times 365\ \textit{days} \times 100\ \textit{years}}$$

The problem is set up as shown above; we begin with 10^{12} seconds, and since one minute is 60 seconds, we place 60 seconds in the denominator so that it will cancel with seconds in the numerator. Using the same rationale, we convert minutes to hours, hours to days, days to years (using 365 days/year), and years to centuries, as shown. If we set the problem up correctly, after cancelling seconds, minutes, hours, days, and years, we end up with centuries, the dimension that we are looking for.

$$\frac{10^{12} \quad \cancel{s} \quad 1 \quad \cancel{m} \quad 1 \quad \cancel{h} \quad 1 \quad \cancel{d} \quad 1 \quad \cancel{y} \quad 1 \quad \text{century}}{60 \quad \cancel{s} \quad 60 \quad \cancel{m} \quad 24 \quad \cancel{h} \quad 365 \quad \cancel{d} \quad 100 \quad \cancel{y}}$$

Cancelling dimensions that appear in both numerator and denominator as shown, the problem becomes:

$$\frac{10^{12}}{60 \times 60 \times 24 \times 365 \times 100}$$

$\approx$ 317.1 centuries (more than 31,700 years).

For perspective, the beginning of recorded history was about 55 centuries ago.

●

THE MATH TUTOR'S FRIEND
Dimension Canceling
Work
Percent
Time
It's easy !!
Speed
Mixtures
Distance
Flow
Mr. Math
WORD PROBLEMS

5. WORD PROBLEMS
TABLE OF CONTENTS

(continues)

5. WORD PROBLEMS
TABLE OF CONTENTS (CONT'D)

•

5. WORD PROBLEMS

Solving "real-world" problems

Word problems are the touted purpose for studying math. After all, we need to solve "real-world" problems, right? There are plenty of real-world problems to solve in every field of endeavor: science, engineering, sales, business, military, surveying, and on and on. We shall discuss only a few of the many possible types of word problems, namely, *percentages*, *time-speed-distance problems*, *work problems*, *stoichiometry problems*, *flow problems*, *volume problems*, and *area problems*.

Given a word problem, how do you proceed? Here's a suggested approach:

(1) *Put down your pencil* and set your calculator aside, then read the *entire* problem statement. This is the most difficult step!

(2) *Decide what kind of problem* it is you are being asked to solve...

(3) *Find out what is needed*; what is the thing you are asked to find? What data are provided? If the variable is not specified in the problem (e.g., "find the width, w"), give it a variable name: d (for distance, or perhaps depth), t (for time), r (for rate, or perhaps radius), A (for area), etc., or just x. Are you being asked for an *exact solution*, or an *approximate solution* (if so, what accuracy is specified)?

(4) Now *read the problem again* and write your formulas, equations, or whatever. In other words, set up the problem. If the units for the data are not consistent, make the needed conversions - you don't want to mix inches and feet, for example. Use dimension analysis to set up your solution. Determine the applicable formula(s) and write them down for reference.

(5) *Do the work.* Don't rely on the calculator alone; use paper and make sketches; write neatly. A good habit is to justify each step of your solution.

(6) *Check your results*! Is the answer reasonable? Is the answer expressed in the specified or proper units (e.g., if the problem data are expressed in feet, then your answer should be in feet or square feet or cubic feet, etc., unless specified otherwise in the problem statement).

Exact *vs* Approximate solutions

Most real-world calculations yield *approximate solutions*. If you took a measurement then you will have an approximate solution and so should specify the accuracy of the measurements. In that case, the answer will (should) be something like "26 ounces, ± 0.5 oz."

When you guess a number between two tick-marks ("| |") on a scale you are *interpolating* (for example, if the left tick-mark is labeled as 8 and the right tick-mark is labeled as 10, and the needle indicator is about half-way between, your guess is 9; if it's a little closer to the right mark, you may use 9 ½ as an approximation.

So, what is an "exact" solution? You will likely be asked to provide an *exact solution* to a problem in a quiz. The short answer is "if you didn't use your calculator, then you probably have an exact solution." Exact solutions include such terms as π, $ln(3)$, $\sqrt{2}$, transcendental numbers, physical constants such as g (for gravity), c (speed of light in a vacuum), and numerous others. For example, a solution of 7.697 is just an approximation (to three decimal places) of the exact solution $2\pi + \sqrt{2}$.

Percent problems

What is "percent?" You can't do percent problems if you're not clear as to what percent (or percentage) is. So, here's a quick review.

> A *percent* is simply part of 100 parts. "1" is one part of a hundred or 1%; 50 is 50 parts of a hundred or 50% (that's half); 100 is 100 parts of a hundred (everything) or 100%. What if you see something like 200% or 1,000% ? Well, 200 percent of something is twice that amount (60 is 200% of 30); 1,000% ten time the amount (50 is one-thousand percent of 5).

In summary, a percent is 1/100 or 0.01. Twenty-five percent of A (25% of A) is $(25\times A)/100$ or $0.25\times A$ or $A/4$.

Convert a fraction or decimal to a percent and vice-versa

To convert a fraction to a decimal, divide the numerator by the denominator (remember, a fraction represents division) then multiply by 100 and append ("tack on") the "%" sign.

Example: $\frac{1}{8}$ is $(0.125)\cdot 100\% = 12.5\%$

To convert any number to a percent, just multiply by 100 and append the "%" sign.

Example: $0.45 = 45\%$; $24 = 2400\%$; $1 = 100\%$

To convert a percentage to a number

Given any percent, convert it to a number; just divide by 100.

Example: $28\% = \frac{28}{100} = 0.28$

Types of Percentage problems

Many basic percent problems relate to the simple proportion

$$\boldsymbol{p}/100 = \boldsymbol{x}/\boldsymbol{y} \qquad (5\text{-}1)$$

where $\boldsymbol{p}$ is a percent expressed with "%" and $\boldsymbol{x}, \boldsymbol{y}$ are values expressed as a ratio (fraction): Given a problem, just solve Equation (5-1) for the unknown value.

Examples:

1) What is 25% of 64?

The proportion is $25\%/100 = \boldsymbol{x}/64$

Solve for $\boldsymbol{x}$: $\boldsymbol{x} = 25\times\frac{64}{100} = 16$

so, 16 is 25% of 64.

2) 30 is what percent of 150?

The proportion is $p/100 = {}^{30}\!/_{150}$

Solve for p: $p = {}^{3000}\!/_{150} = 20\%$ so, 30 is 20% of 150.

3) 60 percent of what is 45?

The proportion is ${}^{60}\!/_{100} = 45/y$

Solve for y : $60y = 4500$

$y = {}^{4500}\!/_{60} = 75$ so, 45 is 60% of 75.

Other percentage-type problems can be posed as follows:

a) An item costs C and sells for S. If S is increased (or decreased) to X, what is the percent increase (or decrease)?

Example: An item costs \$134 and sells for \$175. If the selling price is increased to \$200, what is the percent increase of profit?

Answer: Let S_1 be the original selling price, and S_2 the new selling price. The original profit is $P_1 = S_1 - C = 175 - 134 = \$41$;

the new profit is $P_2 = S_2 - C = 200 - 134 = \$66$.

$$\%\text{ increase} = \frac{P_2 - P_1}{P_1} = \frac{66-41}{41} = \frac{25}{41} \approx 37\% .$$

b) A price P includes x percent tax. What is the actual price less tax (i.e., not counting the tax)?

Example: Assume a 7% sales tax. What is the price, P, of a taxable item sold for \$100.00, tax included?

Answer: $P + 0.07P = \$100$. $P = \$100/1.07 = \$93.46$.

Annual percentage rate (APR)

A word about annual *percentage rate*, or *APR*., which has several definitions and methods of calculation. Therefore, one cannot (or should not) compare APRs when shopping for mortgage loans, say, because different fees may be involved in the calculation.

Basic word problems

These problems ask to state mathematically what is stated in words. Not common in everyday life, the problems are useful for encouraging translation of words to mathematical expressions. They are generally elementary-school type questions.

Example: Three times a number reduced by 5 is equal to 7 plus the number.
Answer: First, identify the unknown number as, say, n. Now write the statement again: $3\times n - 5 = 7 + n$ from which n is easily determined to be 6.

Example: What is the difference between (or "of") 8 and x?
Answer: *Difference* means *subtract* in word problems. The terms of subtraction are *minuend*, *subtrahend*, and *remainder* ("take away" the subtrahend from the minuend to get the remainder). Here, the minuend is 8 and the subtrahend x, so the remainder is expressed as $8 - x$.

Example: Joe is older than Garry by 22 years; the sum of their ages is 158. What are their ages?
Answer: J – G = 22 and J + G = 158, from which J = G + 22. Add the two equations to get 2J = 180 from which J = 90. G = J – 22= 90 – 22 = 68.

Ratios and proportions

Ratios are fractions, so the phrase "*a* is to *b*" or "*a*:*b*" is simply the fraction *a/b*. The key word here is "to."

Example: Express the ratio of 15 less than *x* to 5 more than *x*.

Answer: All the stuff to the left of "*to*" forms the numerator of the ratio and stuff to the right of "*to*" forms the denominator, so the ratio is expressed as $(x - 15)/(x + 5)$.

Proportions are equations. The phrase "*c* is to *d* as *a* is to *b*" means $a/b = c/d$ and by *cross-multiplying* we get $ad = bc$. If we know any three, we can solve for the fourth variable. The variables *a* and *d* are *extremes* in the proportion and the variables *b* and *c* are the *means*.

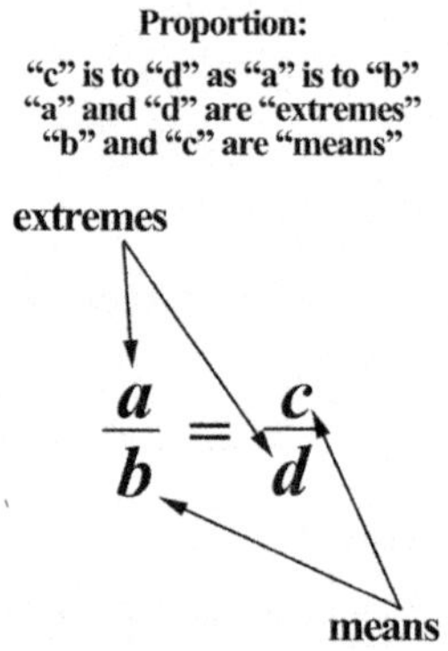

There are many kinds of word problems where you must translate phrases into algebraic expressions, much more than can be covered here. Just take your time and use common sense.

Perimeter, area, and volume problems

Formulas for perimeter and area are (or should be) well known to you by now. But, let's review.

Plane figures

Plane figures are shapes in two dimensions. Here we deal with closed plane figures and deal with perimeter and area.

Perimeter

Think of *perimeter* as a "walk-around." Perimeters are always in linear measurement, e.g., inches, feet, meters, miles, etc… A possible issue for perimeter problems is consistency of units. If the problem specifies measurements in, say, inches for width and feet for length, then either convert the inches to feet (and the answer will be in feet) or convert feet to inches (the answer will be in inches). Check the problem statement; are you asked to express your answer in feet or inches? Unless otherwise specified, then, for a foot-inch problem, if the answer exceeds 12 inches, then your answer should be expressed in feet and inches (e.g., 11' 8"). See the example to the left.

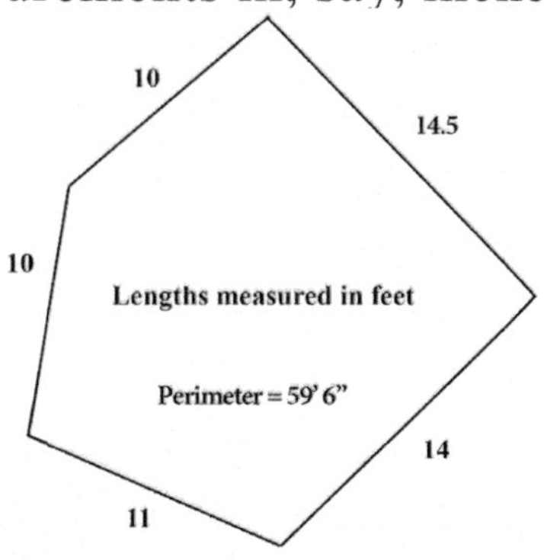

Area

Think of *area* as the amount you must paint (or mow). Area is expressed in *square units*: square inches, square feet, square miles, etc., and for squares and rectangles is the result of multiplying the width by the length of the figure. Of course in the case of a square, the length and width are equal so a square with side of x will have perimeter $4x$ and area x^2 (hence the term "square").

Quadrilaterals

These are four-sided figures including parallelograms (includes squares and rectangles) and the *trapezoid,* also called "trapezium." The area of a trapezoid is given by $(a + b) \times h/2$ where h is the altitude.

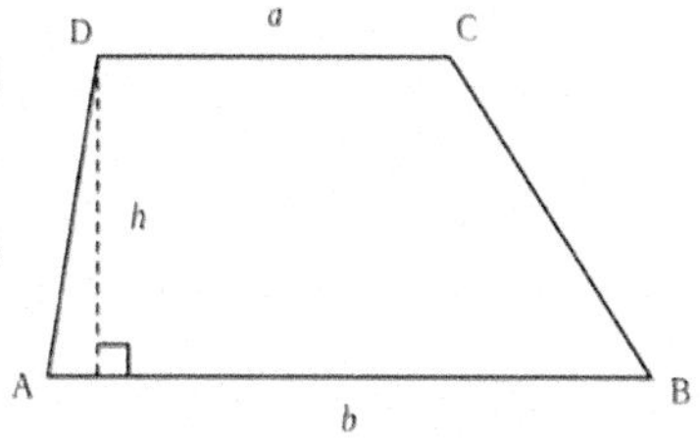

Triangles

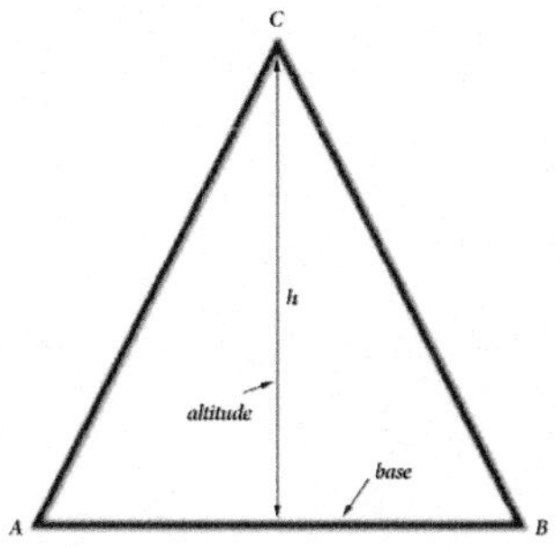

The area of a triangle is given by

$$(base \times altitude)/2$$

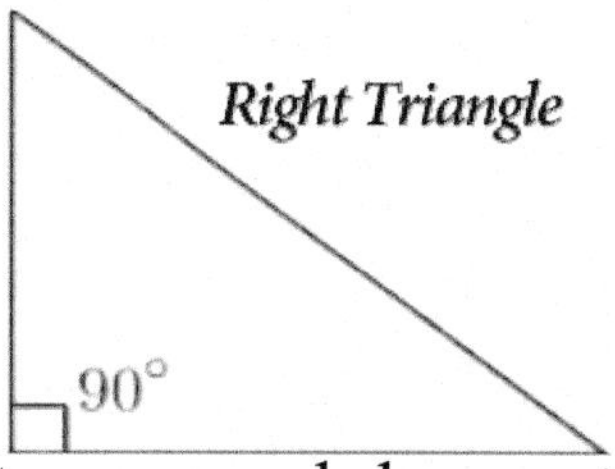

In the case of right triangles, the altitude, h, can be either of the two sides adjacent to the 90° angle. The "long" side, of course, is the hypotenuse, and does not appear in the above formula.

Circles

Refer to the figure to the right. The "perimeter" of a circle is called *circumference* ("*C*"). The radius, "*r*," is the distance from its center ("*O*" for "origin") to the circumference. The diameter, d (not shown) is twice the radius; that is, $d = 2r$. The area of a circle is $A = \pi d = 2\pi r$.

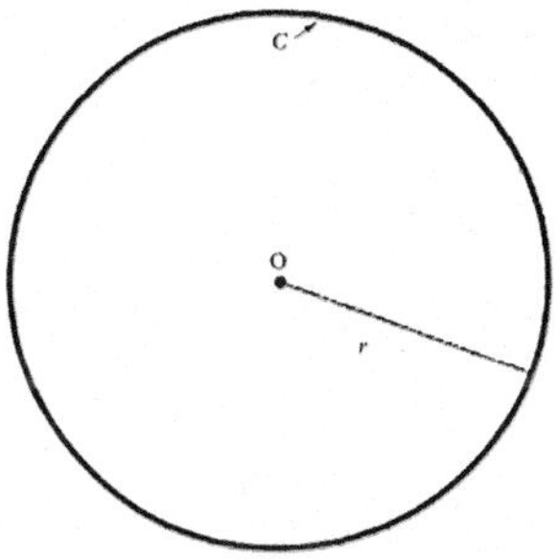

Solids

Volume

Take any container such as a can, box, or cone and fill it with sand or a liquid. The amount of fill then is the *volume* of the can, box, or cone, and is measured in cubic units.

Parallelepiped

Informally, the *parallelepiped* is a "box" in which opposite faces are parallelograms. Consider the *cube*, a solid with six identical square faces (think of a die from a pair of dice). A cube with one edge (i.e., side of one of the faces) that is x units in length will have volume x^3 cubic units (hence "cube"). Quantities such as pint, quart, gallon, and liter are measures of volume, and can be expressed in cubic inches, cubic feet, cubic centimeters, etc. The *surface area* is the sum of the areas of all the faces Each face is congruent (same as) its opposite face.

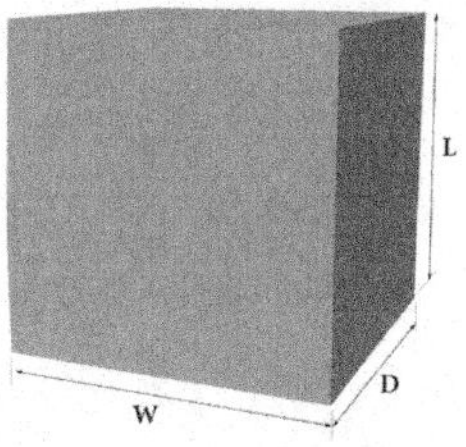

The volume is LWH.

Other solids

There are solids which have other than rectangular faces. It's easy to see that solids can take on an infinite number of shapes. How do you suppose one could estimate the volume of a very complicated solid? One way is to fill it with a liquid and pour the liquid out into a can or box, then calculate the volume by measurement.

The volumes of the following familiar solids may be expressed by formula and some involve the transcendental number π. Most algebra quizzes will provide these and other formulas so there is little need to memorize them.

Cylinder

The volume of a cylinder is $V = \pi r^2 h$

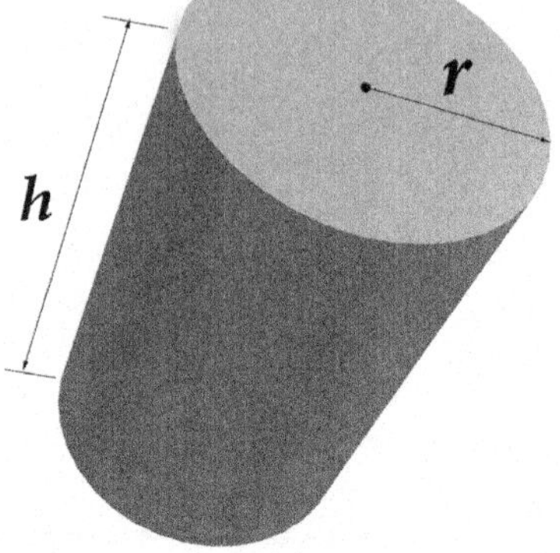

The surface area is $S = 2\pi rh + 2\pi r^2$
(the second term is the sum of the areas of the top and bottom). Now envision that the cylinder is an aluminum can. Remove the top and bottom and then cut the can from top to bottom and open it flat. The result is a rectangle having width h and length $2\pi r$, the original circumference. Multiply these together to get the first term of the surface area. The figure shows a *right circular cylinder.* That is the bases (top and bottom) are circular and the axis joining the two centers of the bases is perpendicular (i.e., at right angles) to the planes of the two bases

Cone

Volume of a cone: $V = (1/3) \times \pi r^2 h$

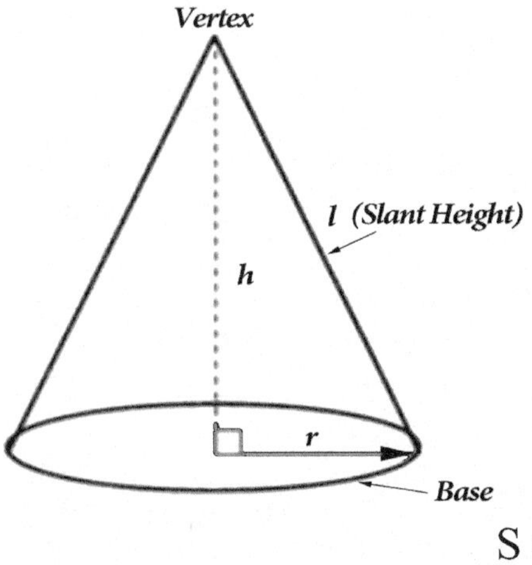

The *volume of a cone* is one-third the area of its base times its height. It is one-third the volume of a cylinder. A line on the surface from the vertex to the base is called the *slant height*. The surface area, S, is given by

$$S = \pi r^2 + \pi rl$$

In a *right circular cone* (axis perpendicular to its base) we can use the Pythagorean theorem to calculate l because it is the hypotenuse of a right triangle with sides r and h:

$$S = \pi r^2 + \pi r\sqrt{(r^2 + h^2)}$$

Mixture problems

These are also called *stoichiometric* problems because of their use in chemistry. *Mixture problems* involve combining different things. If you combine, say, two different mixtures, how much of each thing will be in the combined mixture? For example, if you have two piles of coins and each pile contains half its amount in quarters and if you combine the two piles, half of the new pile will be in quarters also. Makes sense, doesn't it? The fundamental relation has many applications and looks like this:

Sum of individual concentrations × volumes = total concentration × total volume:

$$c_1 V_1 + c_2 V_2 + \ldots + c_n V_n = c_{Total} V_{Total}$$

where the cs are the concentration in percentage and the Vs are the volumes.

Theater tickets

Your local players theater is selling tickets for a new show. The prices are \$15.00 for adults and \$10.00 for "senior citizens" (no children permitted). The theater has 150 seats and you expect a full house. Sure enough, 150 tickets have been sold for a total take of \$2,235, ignoring sales tax. However, you forgot to count the number of senior citizens who have purchased tickets. You do know, however, that because it's a full house, if an "A" amount of tickets were sold to adults, then the remaining tickets sold to senior citizens has to be 150 – A. Set up a table as shown below.

Type Ticket	# Sold a	Unit Price \$$b$	Extension $(a \times b)$
Senior Citizens	150 – A	\$10	\$10(150 – A)
Adults	A	\$15	\$15A
Totals	150		\$2,235

Note that the "Extension" column reveals the equation to be solved:

$$10(150 - A) + 15A = 2235$$
$$1500 - 10A + 15A = 2235$$
$$5A = 2235 - 1500$$
$$5A = 735$$
$$A = 147$$

A total of 147 tickets were sold to adults, so 150 – 147 = 3 tickets were sold to senior citizens. Checking, we see that 3×10 + 147×15 = 30 + 2,205 = 2,235. Problem solved! Here's the table again with the calculated values.

Type Ticket	Sold	Unit Price	Extension
Senior Citizens	3	\$10	\$30
Adults	147	\$15	\$2,205
Totals	150		\$2,235

What has all of this to do with "mixture" problems? Let's analyze the table: $^3/_{150}$, or 2%, of the tickets were for senior citizens and $^{147}/_{150}$, or 98% of the tickets were for adults.

Other Types of Mixture Problems

Liquid solution mixtures

How many ounces of distilled water do you add to 64 ounces of a 12% saline solution to make a saline solution of 10% ? The basic formula is $\boldsymbol{q} = \boldsymbol{pa}$ where $\boldsymbol{q}$ is the quantity of the substance in the solution, $\boldsymbol{p}$ is the percent of the substance, and $\boldsymbol{a}$ is the amount of the solution. Let's set up a table as we did for the tickets problem. Let x be the amount of distilled water needed. There are three types of liquid in this problem: distilled water, a 12% saline solution, and a 10% saline solution. We must solve for x, the amount of distilled water needed in order to make a 10% saline solution from 64 ounces of a 12% saline solution.

Type Liquid	a (amount of liquid)	p (% salt)	$=$ q (amount of salt)
distilled water	x	0	0
12% sol'n	64	12%	$64 \times 0.12 = 7.68$
10% sol'n	$64 + x$	10%	$0.1 \times (64 + x) = 72.8$

As in the ticket problem, use the last column to set up the equation:

$$0 + 7.68 = 0.1 \times (64 + x)$$
$$7.68 = 6.4 + 0.1x$$
$$0.1x = 7.68 - 6.4 = 1.28$$
$$x = 1.28/0.1 = 12.8$$

substituting the value of x the solution is $0.1 \times (64 + 12.8) = 0.1 \times 76.8 = 7.68$. So, 7.68 ounces of distilled water must be added to a 12% saline solution to make 72.8 ounces of a 10% saline solution.

Two more mixture problems

The "table" approach can be used for other kinds of mixtures. Here's a "nutty" problem: How many pounds of Brazil nuts costing \$16 per pound must be mixed with 10 pounds of cashews that cost \$12 per pound in order to make a mixture of nuts that cost \$14 per pound?

	am't (lbs)	unit cost	extended cost (\$)
Brazil nuts	x	\$16	$16x$
cashews	10	\$12	$10 \times 12 = \$120$
mixture	$x + 10$	\$14	$\$14(x + 10)$

Again, from the last column you get the equation:

$$16x + 120 = 14(x + 10)$$
$$16x - 14x = 140 - 120$$
$$2x = 20$$
$$x = 10$$

An equal amount of Brazil nuts and cashews at the specified prices will cost the average (mean) of \$16 and \$12 or \$14.

Spiked punch

Three gallons of punch is mixed with one quart of 80-proof vodka. What is the proof of the punch?

	quarts of punch	*a* alcohol %	*b* quarts of alcohol	proof $2\times a$
punch	12	0	0	0
vodka	1	40%	0.4	80
mixture	13	x	$13x$	$2x$

This time from column *b* we have:

$$0 + 0.4 = 13x$$

$$x = {}^{0.4}\!/_{13} = 0.03 \text{ (2 decimal places)}$$

The spiked punch is about 6 proof. The moral is that before charging off to your calculator, set up the problem in a table.

Mixture ratios

It is possible to equate the ratio of the quantities of a cheaper item to an expensive item using a ratio of costs as follows:

let

- C_C be the unit cost of the cheaper item (usually known)
- C_E be the unit cost of the expensive item (usually known)
- C_M be the unit cost of the mixture (usually specified) where $C_C < C_M < C_E$
- x_C the amount of the cheaper item
- x_E the amount of the expensive item
- x_M the amount of the mixture

Clearly,

$$x_M = x_C + x_E \qquad (5\text{-}1)$$

The total cost of the mixture is

$$C_M x_M = C_C x_C + C_E x_E \qquad (5\text{-}2)$$

Substitute $x_C + x_E$ from Equation (5-1) for x_M in Equation (5-2):

$$C_M(x_C + x_E) = C_C x_C + C_E x_E$$

Clear the parentheses

$$C_M x_C + C_M x_E = C_C x_C + C_E x_E$$

Collect *like* terms

$$C_M x_C - C_C x_C = C_E x_E - C_M x_E$$

Extract factors

$$x_C (C_M - C_C) = x_E (C_E - C_M) \qquad (5\text{-}3)$$

The ratio of cheaper to expensive items is

$$x_C / x_E = (C_E - C_M)/(C_M - C_C) \qquad (5\text{-}4)$$

Summarize the above in a table:

	unit cost	amount	total
cheaper items	C_C	x_C	$C_C x_C$
expensive items	C_E	x_E	$C_E x_E$
mixture	C_M	$x_M\ (= x_C + x_E)$	$C_M x_M = (C_C x_C + C_E x_E)$

Mixture / profit: Watered-down maple syrup

A vendor plans to mix pure maple syrup (costing \$135.00 per 2.5-gallon container) with 40 gallons of sugared water (costing \$0.50/gallon) such that the cost of the mixture will be \$40.00/gallon. The retail price is to be 20% over the cost of the maple syrup. The mixture will be sold in ½-gallon bottles that must be purchased in 12-bottle lots costing \$6.00/lot. Labels (already on hand) are to show the percentage of pure maple syrup in the bottle. The mixing technician costs \$75.00/hour for the four-hour mixing process. FICA taxes of 6.2% must be withheld from the wages and matched by the employer. A 7% sales tax is added to the selling price to consumers. Assume that all bottles of the mixture will be sold. Round to one decimal point; round dollars to nearest cent.

From Equation (5-4), the ratio of water (cheap) to syrup (expensive) is given by

$$\frac{X_W}{X_S} = \frac{C_S - C_M}{C_M - C_W}$$

where

X_W is the amount of water
X_S is the amount of syrup
C_W is the unit cost of water
C_S is the unit cost of syrup

(1) How much syrup is needed?
(2) How many lots of ½-gallon bottles must be purchased?
(3) What is the retail price of the mixture? (less tax)
(4) What is the total cost for the 12-bottle lots?
(5) How many half-gallon bottles will be filled?
(6) How much sales tax is to be sent to the taxing authorities?
(7) What is the total labor cost?
(8) How much FICA tax is collected?
(9) What is the % maple syrup in the mixture?

(10) What is the cost-per-bottle of the mixture?
(11) What is the vendor's net profit?
(12) What is the cost of excess (unbottled) mixture and supplies?
(13) Enter missing data in the table below.

ITEM	UNIT COST/RATE	QUANTITY	EXTENSION
SYRUP	$C_S =$	$X_S =$	
WATER	$C_W = \$0.50$	$X_W = 40$	\$20.00
MIXTURE	$C_M = \$40.00$	$X_M =$	
LABOR	\$75		
FICA	12.4%		
BOTTLES (empty)	\$0.50		
BOTTLES (full)			
SALES TAX	7%		

Solution:

(1) *How much syrup is needed?*

Before applying the proportion formula we must determine C_S, the unit cost of the syrup which costs \$135 for 2.5 gallons:

$$C_S = \$135.00/2.5 = \$54.00/\text{gallon}$$

We have C_M and C_W from the table; now apply the formula:

$$\frac{X_W}{X_S} = \frac{C_S - C_M}{C_M - C_W}$$

$$\frac{40}{X_S} = \frac{54 - 40}{40 - 0.5} = \frac{14}{39.5}$$

$$X_S = \frac{40 \times 39.5}{14} \approx 112.9 \; gallons$$

(2) *How many lots of ½-gallon bottles must be purchased?*
The amount of mixure is the sum of the syrup and water:

$$X_M = X_S + X_W = 112.9 + 40 = 152.9 \text{ gallons.}$$

Containers; 152.9 gallons are equivalent to 305.8 half-gallons. Containers are sold in 12-bottle lots. The lots needed are given by

No. of lots = 305.8/12 = 25.48, so 26 lots will be needed.

(3) *What is the retail price of the mixture? (less tax)*
The retail price is to be 20% over the $54 cost of the syrup:

$$\text{Cost/gallon} = 1.2 \times \$54 = \$64.80\text{/gallon.}$$

(4) *What is the total cost for the 12-bottle lots?*
Twenty-six 12-bottle lots are priced at $6.00/lot

$$\$6 \times 26 = \$156.$$

(5) *How many ½-gallons bottles can be filled?*
The 152.9-gallon mixture amount translates into 305.8 half-gallon units; the bottles must be full, so there will be

$$\text{INT}(305.8) = 305 \text{ full half-gallon bottles}$$

(6) *How much sales tax is to be sent to the taxing authorities?*
Sales tax is 7% × $32.40 = $2.268 rounded to $2.27/bottle.
Assuming all bottles are sold, the total tax will be

$$\$2.27 \times 305 = \$692.35$$

(7) *What is the total labor cost?*
Total labor cost includes wages and the employer's portion of the FICA tax. The wage rate is $75/hour and the employer's share of FICA is 6.2% of wages paid.

$$\text{Wages} = 4 \text{ hrs} \times \$75\text{/hr} = \$300$$
$$\text{Employer's FICA} = 0.062 \times \$300 = \$18.60$$
$$C_{\text{labor}} = \$300 + \$18.60 = \$318.60$$

(8) *How much FICA tax is collected?*
This is the sum of the FICA withheld from wages and the employer's share, or 12.4% of the gross wages:

$$\$300 \times 12.4\% = \$37.20$$

(9) *What is the % maple syrup in the mixture?*
This is computed as follows:

$$(X_S/X_M) \times 100\% = {}^{112.9}\!/_{152.9} \times 100\% = 73.8\%$$

(10) *What is the cost of the mixture per bottle?*
At C_M = $40/gallon (given), each half-gallon bottle costs $20.00.

(11) *What is the cost of excess (unbottled) mixture and supplies?*

$$C_{\text{excess}} = \text{Cost of excess mixture} + \text{Cost of unused bottles}$$

Of the original 152.9 gallons of mixture, 305/2 half-gallon batches = 152.5 gallons was used leaving an excess of 0.4 gallons × $40/gallon = $16.00 value.

There were 26 12-bottle lots = 312 fifty-cent bottles of which 305were used, leaving an excess of 7 bottles valued at $3.50. Therefore,

$$C_{\text{excess}} = \$16.00 + \$3.50 = \$19.50.$$

(12) *What is the vendor's net profit? Percent profit?*

Net Profit = Gross Sales – Total Expenses

= Gross Sales – (Cost of product + Cost of material + Cost of Labor + Cost of excess

$= (305 \times \$32.40) - (C_M + \$156.00 + \$318.60 + \ \$19.50)$

$= \$9882.00 - (6100.00 + 156.00 + 318.60 + 19.50) = (9882 - 6613.60) = \$3{,}268.40$

% profit = [(gross sales – total expenses)/gross sales]×100%

= (9882.00 – 3268.40)/9882.00]×100% ≈ 67%

(13) *Enter missing data in the table below.*

ITEM	*UNIT COST/RATE*	*QUANTITY*	*EXTENSION*
SYRUP	$C_S = \$54.00$	$X_S = 112.9$	$6,096.00
WATER	$C_W = \$0.50$	$X_W = 40$	$20.00
MIXTURE	$C_M = \$40.00$	$X_M = 152.9$	$6,116.00
LABOR	$75.00	4	$300.00
FICA	12.4%	$300	$37.20
BOTTLES (empty)	$0.50	312	$156.00
BOTTLES (full)	$32.40	305	$9,882.00
SALES TAX	7%	$9,882.00	$691.74

Dimension analysis

(also "*dimension cancelling*" - see the previous section for a detailed discussion) Think back to your study of fractions in grammar school. When you divide a number by a fraction, you invert the fraction and then multiply. For example: 20 divided by ⅔ is 20 × (3/2) (= 30). So, divide *x* by 12*x*/*y*: the answer is *xy*/12*x*; the *x*s cancel and you get *y*/12. Now, suppose that we want to convert 3 miles into feet (i.e., no. of feet in three miles). Knowing that 5,280 feet equals one mile, we can write 3 miles divided by 5,280 feet/mile (a fraction). Continuing, 3 miles/5,280 feet/mile ... wait a second, we'll end up with 4 miles×miles/5,280 feet or 3 miles ^{2}/5,280 feet. Something's wrong! We were looking for an answer in feet. So we apply a little dimensional analysis. We are given miles and need the answer in feet, so write down the problem statement and the desired dimension (feet): 3 miles × feet. We must get rid of "miles" so we need "miles" dimension in the denominator to cancel* the "miles" in the numerator, and we know that there are 5,280 feet in a mile, or 5,280 feet/mile. Back to the problem, just write 3 miles × 5,280 feet/mile. The "miles" cancel and the

* The term "cancel" in a fraction really means "multiply by 1." If you divide 5 by 5 the quotient is 1; that is, anything divided by itself is 1, so too, miles/miles = 1 and inches/inches = 1. Finally, 1 times anything doesn't change it, so 1 x feet = feet. This is the secret, or "trick" to dimension canceling is; all you need know is the conversion ratio (12 inches/foot; 4 quarts/gallon; 1,000 milliliters/liter, and so on).

answer is $3 \times 5{,}280$ feet $= 15{,}840$ feet. This is *dimension analysis*, a powerful tool for solving problems like "How many years are a trillion seconds?"

Example: In 2022 the national debt of the U.S. was, let's say, $31.4 trillion.
1) How long will it take to pay back if it's done at $100,000/second?

2) What is the national debt per capita, assuming a U.S. population of 333,000,000?

Solution:

1) How long will the 2022 national debt take to pay back if it's done at $100,000/second?

Let n be the number of payments: $\$100{,}000 \times n = 31.4 \times 10^{12}$ from which

$$n = 31.4 \times 10^{12}/100{,}000 = 31.4 \times 10^{7} \text{ payments}$$

31.4×10^{7} payments/(1 payment/second) $= 31.4 \times 10^{7}$ seconds

Set up the problem to facilitate the use of dimension analysis:

$$\frac{31.4 \times 10^{7} \text{ seconds} \times \text{min} \times \text{hr} \times \text{day} \times \text{year}}{60 \text{ seconds} \times 60 \text{ min} \times 24 \text{ hrs} \times 365 \text{ days}}$$

"Cancel" the seconds, minutes, hours, and days to get

$$31.4 \times 10^{12}/(6 \times 6 \times 24 \times 365) = 9.96 \text{ years, approximately.}$$

So, paying off $31.4 trillion at the rate of $100,000 per second will take around ten years, and this is not counting interest!

2) What is the national debt per capita, assuming a population of 333,000,000?
$\$31.4 \times 10^{12}/(333 \times 10^{6}$ persons) equals approximately $94,294 dollars per person. That's what each and every resident of the U.S. owes on a debt of $31.4 trillion!

Rate problems

"Job" problems

Also called "*work problems*," *job problems* are about *rates*; for example, a number of people having differing rates of work who are working together on a project - painting houses of identical surface area, say. Because we are dealing with mathematical techniques here we'll ignore real-world considerations such as breaks, wages, competition and unions. The important term is "rate," whichimplies time, so we deal with amount of *work done per unit time* (per second, per hour, per day, etc.). In the case of workers, for example, we need to determine each person's (amount of work)/(unit time) like "one job per day."

Example: A painter can paint a house of a given size in eight hours; another painter can do the same job in six hours. How long would it take both painters working together to do the job?

Solution: Set up the problem; call the painters A and B and the job J.
Rate for painter A: (one job/8 hrs), or $J/8 = 0.125$ J/hour. So the rate for painter A is 0.125 of a job per hour or (take the reciprocal) 8 hours per job.

Similarly for painter B: $J/6 = 0.167$ J/hour, which is an approximation due to rounding.

Two ways to proceed in solving this problem are to (1) use the fractions, and (2) use the decimals. We should arrive at the same answer and will employ dimension canceling in the process.

(1) *Use the fractions:* Let T be the total time for the two painters A and B doing the job together: 1 job/(8 hours) + 1 job/(6 hours) = 1 job/(T hours). Add the fractions on the left side of the equation by cross-multiplying and keep track of the dimensions:

$$1\text{ job}/(8\text{ hrs}) + 1\text{ job}/(6\text{ hrs}) = (6\text{ hrs} + 8\text{ hrs})/48\text{ hrs}^2$$

[the denominator is $6\text{ hrs} \times 8\text{ hrs} = 48\text{ hrs}^2$]

"hrs" can be factored in the numerator

$$(6\text{ hrs} + 8\text{ hrs})/48\text{ hrs}^2 = (6 + 8)\text{hrs}/48\text{ hrs}^2 = 14/48\text{ hrs}$$

So far we have 14/48 hrs = Job/T hrs, so we must solve for T which is in the denominator. But we are looking for T to be in the numerator! Just "flip" or invert* the fractions:

T hrs/Job = 48 hrs/14, so $T = 3.4$ hours, approximately, or 3 hrs 24 minutes/Job.

(2) ***Use the decimals.*** Actually, this is not the best way because it violates the "no intermediate calculation" rule at the outset. However, we will proceed to show that the results (in this problem) are identical.

A and B doing the job together: 0.125 job/hr + 0.167 job/hr = 1 job/T hrs. Because the denominators are identical (1 hr) we can simply add on the left side of the equation.

$$0.125\text{ job/hr} + 0.167\text{ job/hr} = (0.125 + 0.167)\text{job/hr}$$
$$= 0.292\text{ job/hr} = \text{job}/T\text{ hrs.}$$

As in method (1) we want "hours" to be in the numerator so we invert the fractions on both sides of the equation:

$$T\text{ hrs/job} = 1\text{ hr}/0.292\text{ job, so } T = 1/0.292 \approx 3.4\text{ hrs as before.}$$

Note: I did not convert the fractions until the very end. Generally it is wise to avoid intermediate rounding, etc. as they it produce rounding errors which become worse as the work progresses. Get to the answer then do your rounding.

Coin problems

Numbers of coin denominations

Example: A sack contains 24 coins. There are twice as many dimes as quarters and three times as many nickels as quarters and twice as many pennies as nickels. How many of each type coin do we have?

Solution: Let q, d, n, and p be the numbers of quarters, dimes, nickels, and pennies, respectively. Reading the problem again, we see that $d = 2q$, $n = 3q$, and $p = 2n$.

* Is this allowed? Let's justify inversion of fractions. That is, we want to show that the equation $a/b = c/d$ is the same as $b/a = d/c$. Cross-multiply the first equation to get $ad = bc$; now cross-multiply the second equation to get $bc = ad$. Clearly, these are the same after cross-multiplication (Law of Symmetry: $x = y$ is same as $y = x$).

Just set up an equation in terms of one kind of denomination and then work out the others. Let's put everything in terms of quarters.

There are q quarters

$2q$ dimes because $d = 2q$

$3q$ nickels because $n = 3q$

$2\times(3q)$ pennies because $p = 2n$ and $n = 3q$.

So, $q + 2q + 3q + 6q = 24$

factoring q: $q(1 + 2 + 3 + 6) = 12\,q = 24$ from which $q = {}^{24}\!/_{12} = 2$.

There are 2 quarters, 4 dimes, 6 nickels, and 12 pennies which total 24 coins, and the problem is solved.

Values of coins and bills

Example: A man opens his wallet and sees that he has only twenty-dollar bills and two-dollar bills (yes, they're still around). The seven bills in the wallet total to \$86.00. How many of each type bill does he have? One way to solve this is to set up a table as shown below. Let x be the number of twenties in the wallet; then the number of twos will be $7 - x$.

Denomination	No. of bills	Value
\$20	x	$20x$
\$2	$7 - x$	$2(7 - x)$
Totals	7	\$86

The last column of the table provides the relevant equation:

$$20x + 2(7 - x) = 86$$
$$20x + 14 - 2x = 86$$
$$20x - 2x = 86 - 4 = 72$$
$$18x = 72$$
$$x = 4$$

There are four twenties and 7– 4 = three two-dollar bills in the wallet.

Flow problems

Example: Three valves can be used to fill a large tank; each valve produces flow at different rates: valve A can fill the tank in 4 hours, valve B can do the job in 8 hours. A variable-rate pump is available and the tank must be filled in 1 hour using the pump and both valves A and B. What should the pump rate be set to in order to do the job?

As with the "job" problem, set up the problem as follows:

$$(1 \text{ tank}/4 \text{ hrs} + 1 \text{ tank}/8 \text{ hrs} + x) = 1 \text{ tank}/2 \text{ hrs}$$

where x is the required rate of the pump.

$$2 \times (\tfrac{1}{4} + \tfrac{1}{8} + x)/\text{hrs} = 1$$

$$\tfrac{2}{4} + \tfrac{2}{8} + 2x = 1$$

From which x is easily computed: x must flow to fill the tank by itself in 8 hours.

Rate, time & distance problems

A review of dimension analysis

The basic measurements, or *dimensions*, are length (L), mass (M), time (T), and electric charge which we do not consider in this book series. That is, all measurable physical quantities are in terms of M, L, and T: the rate of speed is distance/time, or L/T; volume is in linear units cubed, or L^3; area has dimensions L^2, and so on. Let's look at the formula $d = rt$, that is "distance equals rate × time." A dimension analysis shows that this is $L = L/T \times T$. The "Ts" cancel and we are left with L = L, so the dimensions are consistent. So what? Well, had we mistakenly written $r = dt$ which has dimensions $L/T = L \times T$ you can see that something is wrong because dimension analysis (dimension canceling) shows an inconsistency. Dimension analysis is used by engineers and scientists to check consistency of formulas and equations. Note that dimension analysis, while ensuring consistency, does not ensure accuracy or correctness of answers. For example, if you multiply feet by inches to compute area, you will have consistent dimensions (L^2) but a wrong answer.

Rate, time & distance

The relationship among rate, time, and distance is

$$d = rt$$

That is, *distance* equals *rate* times the *time*. If you know or are given any two of these, the third can be calculated.

River swimmer

A person swims across a river with a current of 6 miles per hour. The swimmer's rate is 3.5 miles per hour and the river width is 1,000 feet.

(1) How many feet downstream will the swimmer reach the other shore?

(2) How far has the swimmer traveled? Give the answer to (2) in both feet and miles.

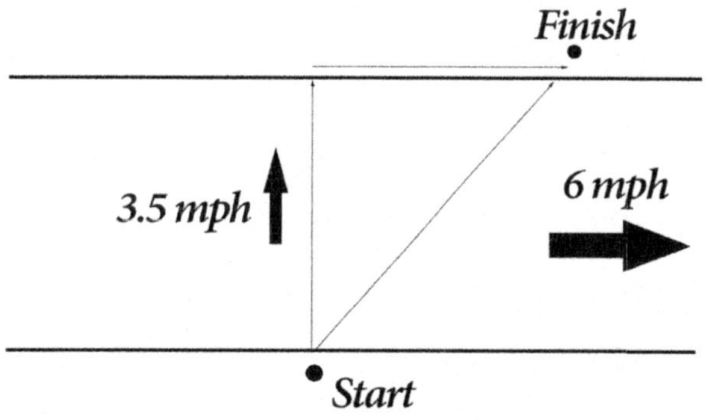

The first thing to do is sketch the problem, as shown above.

Next, set up the problem in consistent units; we could convert miles to feet, but in this case perhaps converting the river width to miles using ratios of miles to feet is simpler.

$$1 \text{ mile}/5{,}280 \text{ feet} = d_S \text{ miles}/1{,}000 \text{ feet}$$

where d_S is the distance the swimmer is to travel (i.e., the width of the river). Cross-multiplying:

$$d_S = {}^{1000}/_{5280} = 0.1894 \text{ or } 0.2 \text{ miles to one decimal place.}$$

(1) The swimmer reaches the opposite shore some distance downstream. Let r_S be the swimmer's rate and d_S be the width of the river. Also, let r_R be the speed of the current, and d_R the distant that the river moved (think of a floating chip) in the same time that it took for the swimmer to reach the opposite shore. So *time* is the common value. Let's express the situation algebraically using the equation $d = rt$. Solve for t [the dimensions are within brackets.]

$$t_S = d_S / r_S \; [\mathrm{T = L/(L/T) = TL/T = T}]$$

and

$$t_R = d_R / r_R$$

time, t, is the same for both equations $t_S = t_R$ so we can write

$$d_S / r_S = d_R / r_R.$$

Plugging in the known quantities we have

$$0.1893 \text{ miles}/3.5 \text{ miles per hour} = d_R \text{ miles}/6 \text{ miles per hour}$$

Good! The only unknown is d_R, the distance downstream.

$$d_R = 6 \text{ miles/hr} \times (0.1894) \text{ miles} / 3.5 \text{ miles/hr}$$

$$[\mathrm{L = L^2/T \times T/L = L^2T/LT = L}]$$

$$= 0.3247 \text{ miles or } 1{,}714 \text{ feet downstream.}$$

(2) How far has the swimmer traveled? Here we appeal to the Pythagorean Theorem and will use distance in feet. The distance that the swimmer traveled is the hypotenuse of a right triangle. The theorem states

$$c^2 = a^2 + b^2$$

where a and b are the lengths of the legs of the triangle (the width of the river and the distance downstream) and c is the length of the hypotenuse (the distance from start to finish), which is what we are seeking.

$$c^2 = d_S^2 + d_R^2$$

$$= (1{,}000)^2 + (1{,}714)^2 = 1{,}000{,}000 + 2{,}937{,}796$$

$$= 3{,}937{,}796 \text{ feet}^2$$

That is, $c^2 = 3{,}937{,}796$ square feet. Taking the positive square root (distance is positive), we have 1,984 feet or 0.38 miles (approximately) traveled by the swimmer.

Overtaking auto problem

At noon, a motorist going at 60 miles per hour passes a "0-mile" milepost at the state line on an interstate highway. Thirty minutes later, another motorist passes the same milepost. The second car overtakes the first at 3:00 p.m. What is the speed of the second car? What will be the nearest milepost when the cars are abreast?

We'll do the second part first because we have all of the data we need: Three hours have elapsed and the rate was 60 miles per hour. Using $d = rt = 60 \times 3 = 180$ miles. The first car had a 30 minute head start so the second car must go 180 miles in 2½ hours in order to be at the 180-milepost at 3:00 p.m. so using $d = rt$ we have $180 = r \times 2.5$ from which $r = 180/2.5 = 72$ miles per hour, the answer we are seeking.

Rotation problems

The diameter of the wheels on a car is 3 feet. The car moves forward and the wheels make 1,000 rotations. How far did the car travel?

Each full rotation of the wheels moves the car a distance equal to the circumference of the wheel, so the car will have traveled 1,000 times the circumference:

$$D = 1{,}000 \times 3\pi \text{ or } 3{,}000\pi \text{ feet (exact solution)}$$

$$D = 1{,}000 \times 3\pi = 3{,}000\pi = 9{,}425 \text{ feet (approximate solution).}$$

●

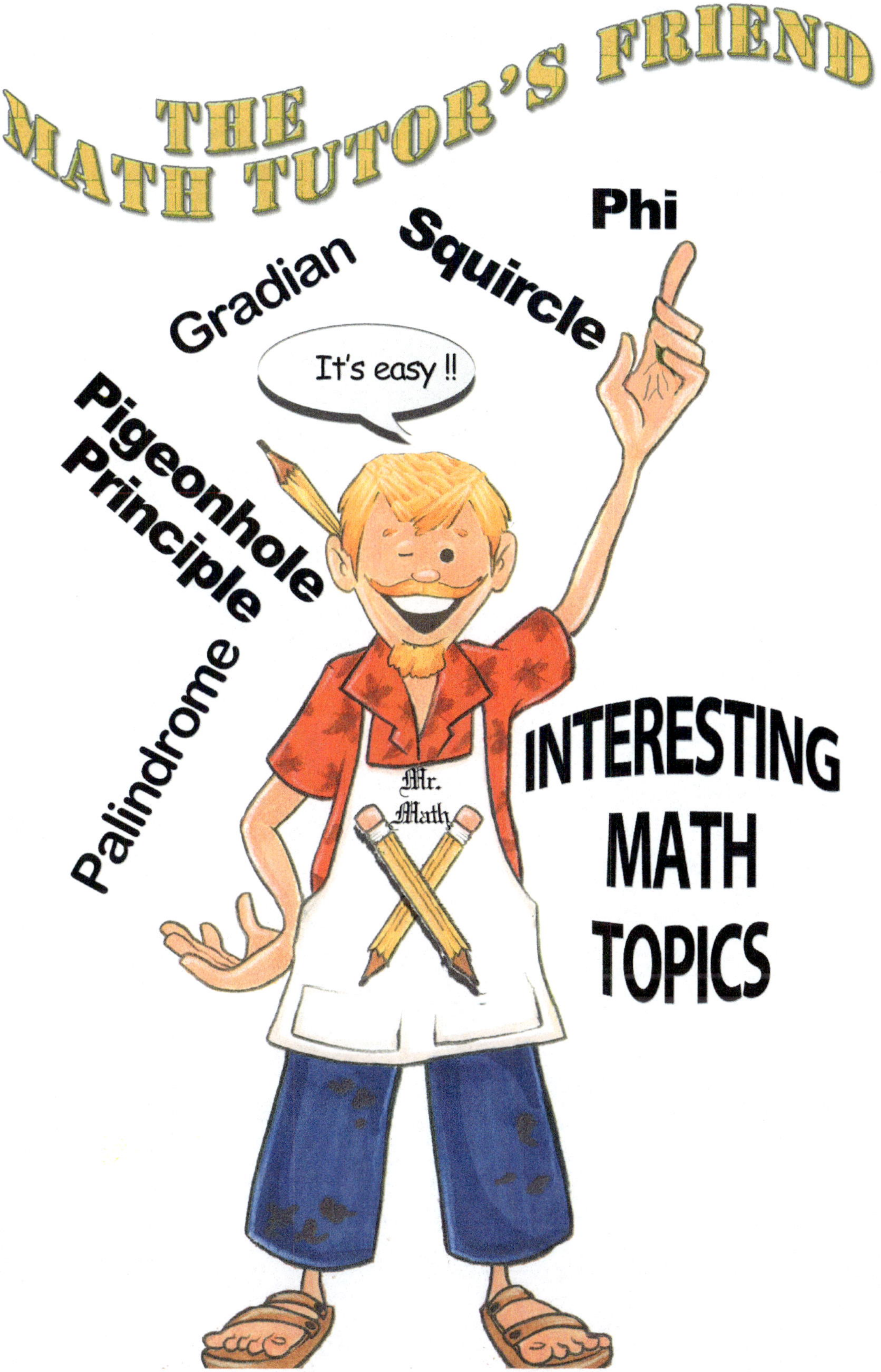
THE MATH TUTOR'S FRIEND
Gradian
Squircle
Phi
It's easy !!
Pigeonhole Principle
Palindrome
Mr. Math
INTERESTING MATH TOPICS

6. INTERESTING MATH TOPICS

TABLE OF CONTENTS

●

6. INTERESTING MATH TOPICS

Degrees, radians, and . . .

Gradian

Here's one for you geometry and trig students. You already know about angular measurement in degrees and radians, right. Well, a third angular measurement unit is the *gradian*, also called *metric degree*, *gon*, and *new degree*. Abbreviations for the gradian include "gr," "grd," and sometimes merely "g" ("g" is also the gravitational constant and is also used in the metric system to represent a gram). There are 400 gradians (grads) in a complete circle (i.e., one *turn*). Each quarter circle (90 degrees or $\pi/2$ radians) is simply 100 gradians. Gradians are used by surveyors and artillery gunners. Many scientific calculators allow the use of degrees, radians, and gradians.

Prime numbers and square roots

A way to determine whether or not a positive integer is *prime* is to divide it by all numbers between 1 and itself. Testing large natural numbers this way certainly seems to be a chore. Luckily there's a shorter way of determining if a given whole number is prime. Consider the *composite* whole number, n; it can be factored as shown in Table 6-1 below for $n = 20$:

FACTORS OF $n = 20$						
Factor 1	1	2	4	5	10	20
Factor 2	20	10	5	4	2	1

Table 6-1

Note the symmetry; the number columns of Table 6-1 contain the pairs of factors of 20; Factor 1 is increasing and the product of each pair (column) is equal to 20. Note also that the 4th, 5th, and 6th pairs are reverse repeats of the first three pairs and so are redundant (if we know the first three then we know the last three), so all that is needed are the first three pairs, which give us all of the factors. Let's try another; this time, in Table 6-2 we'll look at the factors of composite number $n = 36$, which is a perfect square:

FACTORS OF $n = 36$									
Factor 1	1	2	3	4	6	9	12	18	36
Factor 2	36	18	12	9	6	4	3	2	1

Table 6-2

In Table 6-2, the reverse repeating pairs begin in the column after the one having "4" as Factor 1 and we see that $4^2 = 16$, which is less than n ($= 20$). Note that 4 is less than the square root of 20, i.e., $4 < \sqrt{20}$. In the case of the perfect square 36 the reverse repeating pairs begin in the column after the column having 6 as both Factor 1 and Factor 2 and we see that $6^2 = 36$, i.e., $6 = \sqrt{36}$. It appears from the examples above that every composite number has a factor less than or equal to its square root; that is, a number that has no divisor less than its square root must be prime.

As a rule, we do not offer formal proofs of conjectures in this review series, leaving such matters to the textbooks. However, proving the conjecture that every composite number has a proper factor less than or equal to its square root is easy and straightforward, so it appears below. If we can prove that any composite number has at least one prime proper factor less than or equal to that number, then any number for which this is not true must be prime. The *proof is by contradiction*: we make an assumption (i.e., conjecture) then show that it results in a contradiction and thus cannot be true (*reductio ad absurdum*):

> Assume n to be a composite number, then $n = ab$ where a and b are both between 1 and n. If $a > \sqrt{n}$ and $b > \sqrt{n}$ then $a \times b > (\sqrt{n}) \times (\sqrt{n}) = n$, that is, $ab > n$. This contradicts our assumption that $n = ab$. Therefore either a or b (or both) must be less than or equal to $\sqrt{n}$. In other words, if n is composite, it has a prime factor, p, such that $p \leq \sqrt{n}$. If no such p exists, then n must be prime.
>
> Example: Is the integer 491 prime?
> The square root of 491 is 22.16, approximately, so we need determine only if 491 is divisible by primes less than 22, namely, 2, 3, 5, 7, 11, 13, 17, 19. Appeal to the divisibility rules to remove 2 (491 is not even), 3 (sum of digits not divisible by 3), 5 (units digit neither 0 nor 5) and 7 (9 times the units digit subtracted from the remaining digits of the number is not divisible by 7). Therefore, we need only try divisions by 11, 13, 17, and 19 to determine that none of these is a divisor of 491. Consequently, 491 is indeed a prime number.

Geometric mean

The *geometric mean* is a measure of central tendency of a set of positive numbers. It is the nth root of the product of the n numbers. For example, the geometric mean of 2,3,4,5 (the product = 120), is the 4th root of 120 (approximately 3.31). In comparison, the arithmetic mean of 2,3,4,5 is $^{14}/_{4} = 3.5$. Finding the geometric mean of a larger set of numbers such as 2,34,54,3,4,9,15 leaves us with the formidable task of finding the 7th root of the product 5,948,640 (write as 5 948 640). But wait! Let's use logarithms.

$$\text{Let } x = (5\,948\,640)^{1/7}$$

Simply take the log of both sides and employ the law of logarithms:

$$\log x = (1/7)\log(5\,948\,640) = 0.967774$$

from which (using a calculator) $x = 9.28$ to two decimal places. As with the median, the geometric mean is less affected by outliers than is the arithmetic mean.

Palindromes

A *palindrome* is a word, phrase or sentence that reads the same forward or backward; examples: eve; kayak; "madam I'm Adam;" racecar; radar; and so on. Here's a palindromic sentence using words instead of letters as above (thanks to *Wikipedia*): "Is it crazy how saying sentences backwards creates backwards sentences saying how crazy it is?"

Mathematical palindrome

Numbers, too, can be palindromic. For example 525 and 333333 are mathematical palindromes. Pascal's Triangle (left), discussed below, is an example of a set of mathematical palindromes. Note that each row is a palindrome.

```
              1
            1   1
          1   2   1
        1   3   3   1
      1   4   6   4   1
    1   5  10  10   5   1
  1   6  15  20  15   6   1
1   7  21  35  35  21   7   1
```

Square number

The term "*perfect square*" is probably more familiar. Note that the numbers in the diagonal of the multiplication table (1, 4, 9,16, 25, ...) are *square numbers*. Rational numbers for which both numerator and denominator are square are also square numbers. It should be evident that square numbers must be positive (or zero). Perfect squares are mentioned later in a detailed discussion of Pascal's Triangle.

Congruum

A *congruum* (plural *congrua*) is a positive integer which results in a square number when added to or subtracted from a square number; thus is lies between successive square numbers in an arithmetic progression of three squares. Given a square, find a positive integer that, whether added to or subtracted from that square, yields a square. If x^2, y^2, and z^2 (for integers *x, y,* and *z*) are three square numbers that are equally spaced apart from each other, the spacing between them, $z^2 - y^2 = y^2 - x^2$, is called a *congruum*. For example, 24 is a *congruum* because $49 - 25 = 24$ (also $5^2 - 1^2 = 24$). Let ***m*** be a positive integer that is the difference between successive squares: that is, if the squares of integers *x, y,* and *z* (i.e., x^2, y^2, and z^2) are equally spaced by the amount ***m***, then $\boldsymbol{m} = z^2 - y^2 = y^2 - x^2$, is a *congruum*. The first ten *congrua* are:

24, 96, 120, 216, 240, 336, 384, 480, 600, 720 ...

(this sequence is A256418 in the O.E.I.S.)*

Congrua are used in cryptography.

Factorial

A *factorial* of a positive integer *n*, denoted by *n*! is the product of an integer and all the integers below it. That is, a positive integer, *n*, has factorial $n! = n(n-1)(n-2)\ldots(1)$. For example,

$$8! = 8\times7\times6\times5\times4\times3\times2\times1 = 40{,}320$$

By definition, $0! = 1$. Factorials can usually be found directly on modern calculators. Factorials are used in Probability and Statistics problems to find combinations and permutations.

* On-Line Encyclopedia of Integer Sequences

Fibonacci sequence

We know about arithmetic and geometric sequences. Here's an interesting one called the *Fibonacci sequence*:

$$0, 1, 1, 2, 3, 5, 8,13, 21, 34, 55, \ldots$$

O.E.I.S A000045

What is the next number in the sequence? Start with the third number, 1; it is the sum of the two preceding numbers (0 + 1). This is the rule: any number in the sequence is the sum of the two preceding numbers. Therefore, the next number (not shown) must be 34 + 55 = 89. The rule for this sequence can be written as a formula with F_n for the nth Fibonacci number:

$$F_n = F_{n-1} + F_{n-2}, \text{ with } n \geq 3 \text{ (because } F_1 = 1 \text{ and } F_2 = 1).$$

Fibonacci sequences appear in biological settings, such as branching in trees, the arrangement of leaves or flower petals on a stem, the fruit sprouts of a pineapple, the flowering of an artichoke, an uncurling fern and the arrangement of a pine cone's clusters, the shapes of mollusk shells, etc. The Fibonacci sequence is actually embedded in Pascal's Triangle. Wow! There is much more about Pascal's Triangle later in this section.

Interesting symmetries of integers

Books and articles abound on the interesting aspects of math at all levels. Refer to Table 6-3 below; take the number 1, square it to get 1. That's not very interesting, but now take 11 and square it to get 121, a palindrome. Now try squaring 111; you'll get 12321, another palindrome. Look at the symmetry!

1^2	=	1
11^2	=	121
111^2	=	12321
11111^2	=	1234321
11111^2	=	123454321
111111^2	=	12345654321
1111111^2	=	1234567654321
11111111^2	=	123456787654321
111111111^2	=	12345678987654321
1111111111^2	=	1234567890987654321

Table 6-3

The square of a string of more than ten 1s is not a palindrome, however. Here's a way to remember the results of Table 6-3: Count the number of 1s in the given number, say there are 6 of them (111111); the square is given by writing the sequence 123456 followed by the decreasing sequence 54321 to yield 12345654321. That's easy to remember!

Table 6-4 displays an interesting pattern generated as follows: multiply a decreasing sequence by 9 and then add a decreasing number to yield a series of palindromic 8s. A similar pattern of 1s is generated by multiplying an increasing number by 9 and then adding a number in the increasing sequence shown in Table 6-5.

9	×	9	+	7	=	88
98	×	9	+	6	=	888
987	×	9	+	5	=	8888
9876	×	9	+	4	=	88888
98765	×	9	+	3	=	888888
987654	×	9	+	2	=	8888888
9876543	×	9	+	1	=	88888888
98765432	×	9	+	0	=	888888888

Table 6-4

1	×	9	+	2	=	11
12	×	9	+	3	=	111
123	×	9	+	4	=	1111
1234	×	9	+	5	=	11111
12345	×	9	+	6	=	111111
123456	×	9	+	7	=	1111111
1234567	×	9	+	8	=	11111111
12345678	×	9	+	9	=	111111111
123456789	×	9	+	10	=	1111111111

Table 6-5

Finally, look at Table 6-6 to appreciate the symmetric, if not palindromic, relationship shown.

1	×	8	+	1	=	9
12	×	8	+	2	=	98
123	×	8	+	3	=	987
1234	×	8	+	4	=	9876
12345	×	8	+	5	=	98765
123456	×	8	+	6	=	987654
1234567	×	8	+	7	=	9876543
12345678	×	8	+	8	=	98765432
123456789	×	8	+	9	=	987654321

Table 6-6

Pascal's Triangle

The array of numbers shown below is known as Pascal's Triangle; let's see how it is constructed.

```
        1
      1   1
    1   2   1
  1   3   3   1
1   4   6   4   1
```

Constructing Pascal's Triangle

Begin with the digit 1 in the first (top) row, labeled "Row 0." In the row underneath (Row 1), insert two 1s in columns preceding and succeeding the top row entry as shown. Now for Row 2 and subsequent rows, insert a 1 in the columns before and after those for the 1s in the previous row. For each remaining position insert the sum of the two flanking numbers in the previous row. Thus, for Row 2 insert $1 + 1 = 2$ in the center. After inserting 1s in Row 4, insert $1 + 2 = 3$ and $2 + 1 = 3$, as shown. Continue this process for all additional rows to be displayed in the Pascal Triangle. Before proceeding with an analysis of Pascal's Triangle let's describe the concept of the combination of ***n*** things ***k*** at a time denoted by ${}_nC_k$ or $\binom{n}{k}$

As described earlier, the factorial of a positive integer ***n***, denoted by ***n***! is the product of ***n*** and all integers less than ***n***. That is,

$$n! = n\times(n-1)\times(n-2)\times\ldots\times3\times2\times1.$$

Returning to combination of ***n*** things ***k*** at a time, the computation is

$$\binom{n}{k} = \frac{n!}{k!(n-k)!}$$

Analyzing the features of Pascal's Triangle

It turns out that Pascal's Triangle has many remarkable properties, some of which are described below. It will be easier to analyze Pascal's Triangle if we "left-justify" it to display all the columns vertically as shown on the next page. Note that the numbering of rows and columns begins with 0.

Analysis of Pascal's Triangle columns

Column 0 is always "*1*."

Column 1 (*1 2 3 4 ...*) holds the *natural numbers* (shown in boldface).

Column 2 (*1 3 6 10 ...*) holds the so-called *triangular numbers*. You can think of the triangular numbers as the number of dots it takes to make various sized triangles. See the figure to the right. The numbers of gifts reported in the carol "12 Days of Christmas" are triangular numbers.

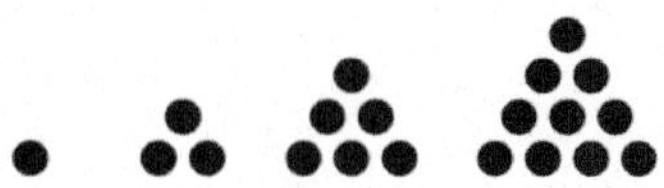

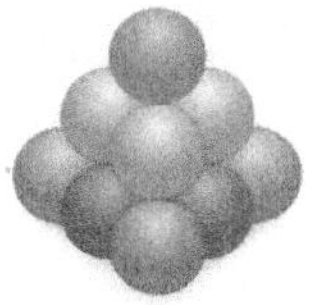

Column 3 (*1 4 10 20 ...*) lists the *tetrahedral numbers*, or *triangular pyramidal numbers*: the number of dots needed to make pyramids with triangle bases.

The columns continue in this way, describing the *simplices* which are just extrapolations of the triangle/tetrahedron idea to higher dimensions. The next column holds the 5-simplex numbers, followed by the 6-simplex numbers and so on. All of this is beyond the scope of this book.

Pascal's Triangle left-justified

Row # \ Col # →	0	1	2	3	4	5	6	7	8	9	10	11	12
0	1												
1	1	**1**											
2	1	**2**	1										
3	1	**3**	3	1									
4	1	**4**	6	4	1								
5	1	**5**	10	10	5	1							
6	1	**6**	15	20	15	6	1						
7	1	**7**	21	35	35	21	7	1					
8	1	**8**	28	56	70	56	28	8	1				
9	1	**9**	36	84	126	126	84	36	9	1			
10	1	**10**	45	120	210	252	210	120	45	10	1		
11	1	**11**	55	165	330	462	462	330	165	55	11	1	
12	1	**12**	66	220	495	792	924	792	495	220	66	12	1

Other features of Pascal's Triangle

Perfect squares

Determine the *perfect squares* of the natural numbers in column 2 by summing the number in the same row of the adjacent column to the right with the number below the number to the right. For example, the column 2 entries 36 + 45 = 81 which is the perfect square of 9.

The rows of Pascal's Triangle

The interesting and remarkable features of Pascal's Triangle described so far may not be of particular value to most algebra students. However, one feature of the rows provides useful information for expanding binomials. Consider

Row #	Col # → 0	1	2	3	4	5	6	7	8	9	10	11	12	Perfect Squares of natural number in Column 1 ↓
0	1													
1	1	**1**												1
2	1	**2**	1											4
3	1	**3**	3	1										9
4	1	**4**	6	4	1									16
5	1	**5**	10	10	5	1								25
6	1	**6**	15	20	15	6	1							36
7	1	**7**	21	35	35	21	7	1						49
8	1	**8**	28	56	70	56	28	8	1					64
9	1	**9**	36	84	126	126	84	36	9	1				81
10	1	**10**	45	120	210	252	210	120	45	10	1			100
11	1	**11**	55	165	330	462	462	330	165	55	11	1		121
12	1	**12**	66	220	495	792	924	792	495	220	66	12	1	

Row # \ Col # →	0	1	2	3	4	5	6	7	8	9	10	11	12
0	1												
1	1	**1**											
2	1	**2**	1										
3	1	**3**	3	1									
4	1	**4**	6	4	1								
5	1	**5**	10	10	5	1							
6	1	**6**	15	20	15	6	1						
7	1	**7**	21	35	35	21	7	1					
8	1	**8**	28	56	70	56	28	8	1				
9	1	**9**	36	84	126	126	84	36	9	1			
10	1	**10**	45	120	210	252	210	120	45	10	1		
11	1	**11**	55	165	330	462	462	330	165	55	11	1	
12	1	**12**	66	220	495	792	924	792	495	220	66	12	1

Pascal's Triangle
left-justified

the binomial $(a+b)^n$, $n \geq 0$ and let's look at row $n = 2$ in the above figure. The Row 2 entries are 1, 2, and 1. Now look at the expansion of the binomial $(a+b)^2$:

$$(a+b)^2 = a^2 + 2ab + b^2.$$

The coefficients of the terms in the expansion are 1, 2, and 1. These are the values in Row 2 of Pascal's Triangle. Note that $(a+b)^0 = 1$ and $(a+b)^1 = a + b$ with coefficients 1 and 1.These are the entries in Row 0 and Row 1 of Pascal's Triangle, as expected. So, we have a useful tool for determining the coefficients of all of the terms in any expansion of a binomial. For example, to expand $(a+b)^7$ simply consult Pascal's Triangle and look at Row 7: (1 7 21 35 35 21 7 1) to write

$$(a+b)^7 = a^7 + 7a^6b + 21a^5b^2 + 35a^4b^3 + 35a^3b^4 + 21a^2b^5 + 7ab^6 + b^7$$

Try to expand $(a+b)^7$ by hand! But wait, we can rewrite the expansion of row 7 as

Here is a method by which one can immediately write the expansion of a binomial to

$$(a+b)^7 = \binom{7}{0}a^7 + \binom{7}{1}a^6b + \binom{7}{2}a^5b^2 + \binom{7}{3}a^4b^3 + \binom{7}{4}a^3b^4 + \binom{7}{5}a^2b^5 + \binom{7}{6}ab^6 + \binom{7}{7}b^7$$

any power ***m*** without having to construct Pascal's Triangle all the way to Row ***m***. This means that every entry in Pascal's Triangle (see the illustration on the next page) is the *combination* of *[Row #]* things *[Column #]* at at time; that is, for row *m*, and the entry in Column *n*,

$$\binom{Row\ \#}{Column\ \#} = \binom{m}{n} \text{ where } 0 \leq n \leq m$$

is the numerical coefficient of the nth term in the expansion of $(a+b)^m$. In addition, the powers of a and b in each term of the expansion are given by $a^{(m-n)}b^n$.

Combinations of [row #] things [column #] things at a time

Row #	Col # →	0	1	2	3	4	5	6	7	8	9	10	11	12
0		1												
1		1	1											
2		1	2	1										
3		1	3	3	1									
4		1	4	6	4	1								
5		1	5	10	10	5	1							
6		1	6	15	20	15	6	1						
7		1	7	21	35	35	21	7	1					
8		1	8	28	56	70	56	28	8	1				
9		1	9	36	84	126	126	84	36	9	1			
10		1	10	45	120	210	252	210	120	45	10	1		
11		1	11	55	165	330	462	462	330	165	55	11	1	
12		1	12	66	220	495	792	924	792	495	220	66	12	1

Powers of base 2

The figure below shows the sum of the rows and also 2 raised to the power of the row numbers; why, they match exactly. How about that?

Row #														Row Sum	2^(Row #)	11^(Row #)
0	1													1	1	1
1	1	1												2	2	11
2	1	2	1											4	4	121
3	1	3	3	1										8	8	1331
4	1	4	6	4	1									16	16	14641
5	1	5	10	10	5	1								32	32	161051
6	1	6	15	20	15	6	1							64	64	1771561
7	1	7	21	35	35	21	7	1						128	128	19487171
8	1	8	28	56	70	56	28	8	1					256	256	
9	1	9	36	84	126	126	84	36	9	1				512	512	
10	1	10	45	120	210	252	210	120	45	10	1			1024	1024	
11	1	11	55	165	330	462	462	330	165	55	11	1		2048	2048	
12	1	12	66	220	495	792	924	792	495	220	66	12	1	4096	4096	

Powers of 11

Remarkably (shown above), the entry of any row of Pascal's Triangle is the value of 11 raised to the power of the corresponding row number.

Sierpinski Triangle

Here's another noteworthy feature: take the original non-justified orientation of Pascal's Triangle; shade in all the odd numbers and you'll get a picture that looks similar to the famous *Sierpinski Triangle* shown on the right.This is one of those fractal designs that goes on forever.

See the Appendix for a summary of he features of Pascal's Triangle.

Fibonacci sequence in Pascal's Triangle

The Fibonacci sequence is also embedded in Pascal's Triangle! Here's the triangle in its left-justified form: Note that the entries in each diagonal sum to a number in the Fibonacci sequence.

Fibonacci sequence

Row #		1	1	2	3	5	8	13	21	34	55	89	144	233
0	1													
1	1	1												
2	1	2	1											
3	1	3	3	1										
4	1	4	6	4	1									
5	1	5	10	10	5	1								
6	1	6	15	20	15	6	1							
7	1	7	21	35	35	21	7	1						
8	1	8	28	56	70	56	28	8	1					
9	1	9	36	84	126	126	84	36	9	1				
10	1	10	45	120	210	252	210	120	45	10	1			
11	1	11	55	165	330	462	462	330	165	55	11	1		
12	1	12	66	220	495	792	924	792	495	220	66	12	1	

Pigeonhole principle

In mathematics, the *pigeonhole principle* (also *shelf principle* or *box principle*) states that if more items are put into containers than there are containers, then at least one container must hold two or more items. This theorem is exemplified in real life by truisms like "in any group of three gloves there must be at least two left gloves or at least two right gloves." According to the Internet, you are born with about 100,000 scalp hair follicles to start with, but it varies with natural hair color (blonds have the most, redheads the least). So let's assume that the the average adult and child has a head that boasts 100,000 hairs. Consider the small city of Waterbury, Connecticut, with a (2005) population of 107,902. According to the pigeonhole principle there were at least two residents who had exactly the same number of hairs on their heads in 2005. Of course, the question of the identities of the iso-hirsute individuals is not answered by this principle.

Squircle

In a Cartesian coordinate system, the circle centered on the point (a, b) with axes parallel to the coordinate axes is described by the equation:

$$(x-a)^2+(y-b)^2=r^2$$

A *squircle* (square circle) centered on the point (a, b) with axes parallel to the coordinate axes is described by the equation:

$$(x-a)^4+(y-b)^4=r^4$$

The figure to the left shows the shape. Squircles are useful in the field of optics. Squircles have also been used in the design of dinner plates. A squircular plate has a larger area (and can thus hold more food) than a circular one with the same radius, but still occupies the same amount of space in a rectangular or square cupboard. The same is true of a square plate, but there are various problems (such as wiping up sauce) associated with the corners of square plates. Button pads on some media players and touch pads on some cell phones are squircles.

Phun with phi

Phi (rhymes with "hi"), with symbol φ, is the 21st letter of the Greek alphabet. One of its uses in mathematics is as the symbol for the "Golden Ratio," defined as the positive solution to the quadratic equation

$$\varphi^2 - \varphi - 1 = 0:$$
$$\varphi = (1 + \sqrt{5})/2$$
$$\varphi \approx 1.6180$$

Two interesting properties of φ:

$$\varphi^3 = \varphi^2 + 1 = (1/\varphi) + \varphi$$
and
$$\varphi^3 = \varphi^2 + 1 = (\varphi + 1)/\varphi\ .$$

Another note of interest is that dividing two adjacent numbers in the Fibonacci sequence (the smaller number is the divisor) yields a quotient that approaches φ as the Fibonacci numbers become larger.

Golden ratio

Two numbers, x and y, are in the golden ratio, φ, if

$$x/(x + y) = y/x.$$

The golden ratio has been known for thousands of years. Phi was described by Johannes Kepler as one of the "two great treasures of geometry;" the other is the Pythagorean Theorem. Phi can be related to Pi through trigonometric functions (angle in radians).

And here's one for you; φ, π, and the first six numbers in the Fibonacci sequence are related as follows:

$$\varphi = 1 - 2cos(3\pi/5).$$

The "0" is implied and the first "1" of the series is the coefficient of φ:

$$0 + 1 \times \varphi = 1 - 2cos(3\pi/5).$$

Great Pyramid of Giza

There is a conjecture that the Egyptians wanted the surface area of one side of the Great Pyramid of Giza (with height h and apothem r) to be equal to the area of a square with side h. Assuming this is so, let's see what happens. The *slant height, s,* is the altitude of a triangular face. Therefore, the area of a face is

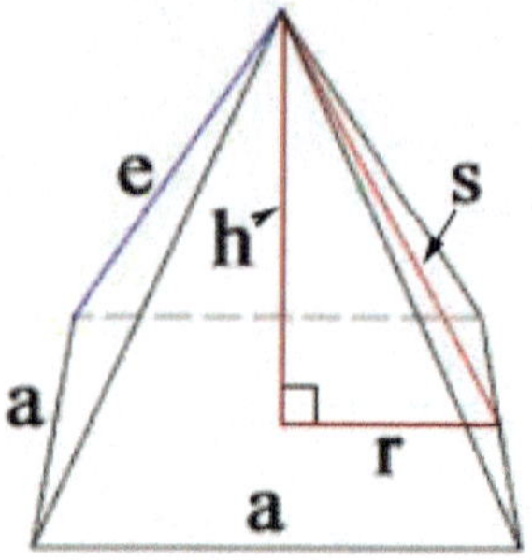

$$h^2 = as/2 \qquad (6\text{-}1)$$

but $a = 2r$, so equation (6-1) becomes

$$h^2 = 2rs/2 = rs \quad (6\text{-}2)$$

From the Pythagorean Theorem we know that

$$h^2 = s^2 - r^2$$

and so equation (6-2) becomes

$$s^2 - r^2 = rs \qquad (6\text{-}3)$$

Subtracting rs from both sides of equation (6-3) yields

$$s^2 - rs - r^2 = 0 \quad (6\text{-}4)$$

which is a quadratic in s with solution

$$s = r(1+\sqrt{5})/2 = r\varphi \approx 1.618r,$$

to three decimal places, and that, my friend, is φ times r. Work out the solution to equation (6-4); a quadratic equation, of course, has two solutions; you will find that only one of the solutions is positive (because length is a positive number). Now, let $r = 1$ and let's do a bit of trig to find the angle, call it θ, of the face of the pyramid and the base. Using the basic trig ratio *cosine* = *adjacent*/*hypotenuse* = $1/s$, we have

$$\theta = arccos(1/s) = \cos^{-1}(1/\varphi) \approx 51.8°.$$

(NOVA ONLINE reports that the actual angle measures 51.5°.)

Other interesting numbers

Perfect number

(No, not "ten")

A *perfect number*, say *n,* is a positive integer that is equal to the sum of its proper divisors (i.e., positive divisor of n other than n itself). The number 6 is a perfect number because its proper divisors (1, 2, and 3) add to 6 (the number 6 is an *improper divisor*, excluded by the definition). The first four perfect numbers are 6, 28, 496, and 8,128. Note that these numbers are all even.* The existence of a perfect number that is odd is an unsolved problem in mathematics. Find one and you'll become famous!

* Every even perfect number can be expressed as $2^{n-1} \times 2^{n}$, where 2^{n-1} is a Mersenne prime (a prime number whose exponent is also prime).

A most beautiful equation
Euler's formula
One of the most important trigonometric identities comes from the Swiss mathematician *Leonhard Euler* (pronounced "oiler") (1709-1783):

$$e^{i\theta} = cos(\theta) + isin(\theta) = cis(\theta)$$

Let $\theta = \pi$, then the equation above* becomes

$$e^{i\pi} = cos(\pi) + isin(\pi) = -1.$$

Now add 1 to both sides the get the famous *Euler's equation* that relates i, e, π, 1, and 0:

$$e^{i\pi} + 1 = 0$$

where

e is Euler's number, the base of natural logarithms,
i is the imaginary unit, which satisfies $i^2 = -1$, and
π is *pi*, the ratio of the circumference of a circle to its diameter.

Euler's equation is cited as an example of deep mathematical beauty. Three of the basic arithmetic operations occur exactly once: addition, multiplication, and exponentiation. The equation also links five mathematical constants: 0, 1, π, e, i. We used Euler's formula in Section 3 to show that i^i is a real number.

Fractions revisited
Let N/D be a fraction where N is the numerator and D ($D \neq 0$) the denominator. Now define the following:
Proper fraction $N < D$; *improper fraction* $N > D$; *vulgar fraction* (also called *common fraction*) where N and D are integers.** A *unit fraction* has $N = 1$, and therefore is the reciprocal of D. The *Egyptian fraction* is a finite sum of unit fractions (reciprocals of whole numbers), that is, of the form $1/D$; $D > 0$, such that each unit fraction differs from the others.

For example, the fraction ¾ can be expressed by the Egyptian fraction ½ + ¼ .

Egyptian fractions are not necessarily unique; ⅘ can be expressed as

$$\frac{1}{2}+\frac{1}{4}+\frac{1}{20} \text{ and also by } \frac{1}{2}+\frac{1}{5}+\frac{1}{10}$$

Over the centuries, mathematicians have developed various identities and *"greedy" algorithms* to find the Egyptian fraction(s) for a given proper fraction. These may be found by searching the internet, and are addressed in this book. Let's apply reasoning for a couple of simple cases.

* $cis(\theta)$ is commonly pronounced "sis theta."

** Of course, if $N = D$ the fraction is unity.

Consider the proper fraction ⅗; we see that 3 is more than one-half of 5 so ⅗ must be ½ plus something and we can write

$$\frac{3}{5} = \frac{1}{2} + x$$

Solving for *x*:

$$x = \frac{3}{5} - \frac{1}{2} = \frac{(6-5)}{10} = \frac{1}{10}$$

so we have

$$\frac{3}{5} = \frac{1}{2} + \frac{1}{10}$$

which is easily verified. The right-hand side of the equation is an Egyptian fraction equivalent to ⅗.

Let's find an Egyptian fraction for the vulgar fraction $^8/_{11}$. Note that 8 is more than one-half of 11, so can immediately write

$$\frac{8}{11} = \frac{1}{2} + x$$

And so

$$x = \frac{8}{11} - \frac{1}{2} = \frac{(16-11)}{22} = \frac{5}{22}$$

We have

$$\frac{8}{11} = \frac{1}{2} + \frac{5}{22}$$

and now must find the Egyptian fraction for $^5/_{22}$:

A little thought reveals that 5 is more than one-fifth of $^5/_{22}$, so

$$\frac{5}{22} = \frac{1}{5} + y$$

from which

$$y = \frac{(25-22)}{110} = \frac{3}{110}$$

And $^3/_{110}$ is easily partitioned into

$$\frac{1}{110} + \frac{2}{110} = \frac{1}{110} + \frac{1}{55}$$

Putting all of this together yields an Egyptian fraction for $^8/_{11}$

$$\frac{8}{11} = \frac{1}{2} + \frac{1}{5} + \frac{1}{55} + \frac{1}{110}$$

which is easily verified using a calculator to add the divisions (recall that a fraction is a division).

Oblong numbers

Square numbers can be represented as $n \times n$. Any whole number n multiplied by $n + 1$ is said to be an *oblong number* (also *pronic number*), sometimes expressed as $\mathbf{O}_n$. The oblong numbers less than 100 are

$$0, 2, 6, 12, 20, 30, 42, 56, 72, 90, \ldots$$

(OEIS A002378)

which are calculated as $0\times1,\ 1\times2,\ 2\times3,\ 3\times4,\ \ldots\ 9\times10$.

Note that the nth oblong number, $\mathbf{O}_n = n(n + 1)$, is twice that of the nth triangular number, $T_n = n(n + 1)/2$. That is,

$$\mathbf{O}_n = n(n + 1) = 2\times T_n = 2\times n(n + 1)/2$$

Runsums

A *runsum* is the sum of a finite number of consecutive integers. The runsum of integers a through b, inclusive, is given by

$$S(a, b) = a + (a + 1) + \ldots + (b - 1) + b\ .$$

Runsums and Triangular numbers

The triangle of circles above represents the triangular number T_7, the sum of the consecutive integers 1 through 7. That is,

$$T_7 = 1 + 2 + 3 + 4 + 5 + 6 + 7 = 28,$$

and is therefore the runsum S(1, 7). That is, T_7 = S(1, 7). Removing the shaded circles truncates the triangle, resulting in a trapezoid having the runsum S(5, 7) = 5 + 6 + 7.

It should be clear from the figure that S(5, 7) = S(1, 7) − S(1, 4).

Developing a formula for the runsum

Consider the sum of the inegers 20 through 30, shown in the table below. The top row contains the integers 20 through 30 and the bottom row the same integers in reverse order.

20	21	22	23	24	25	26	27	28	29	30
30	29	28	27	26	25	24	23	22	21	20

Note that each column sums to 50 and that there are 11 columns. The totals of all the numbers in each row of the table, then, must be $(50\times11)/2 = 550/2 = 275$.

This is precisely the runsum S(20, 30). Now look at the general case for S(a, b), the sum of n consecutive integers between a and b inclusive:

Col 1	2	...	$n-1$	n	*No. of integers*
a	$a+1$		$b-1$	b	*consecutive integers*
b	$b-1$	...	$a+1$	a	*reverse of above*

A look at the table shows that the total of each column is $a + b$.
In terms of a and b, the number of columns, n, is $n = b - a + 1$.

The total of the column sums is $n\times(a + b) = (b - a + 1) \times (a + b)$.

The runsum is the total of one row, i.e., half of the total of two rows, is

$$n \times (a + b)/2 = [(b - a + 1) \times (a + b)]/2$$

Therefore,

$$S(a, b) = [(b - a + 1) \times (a + b)]/2 .$$

Note that in the special case where $a = 1$ we have the triangular number T_b:

$$T_b = S(1, b) = b(b + 1)/2 .$$

How is *runsum(a, b)* related to the triangular numbers T_a and T_b ?
Now, express the runsum in terms of a and b :

$$runsum(a,b) = \frac{(b-a+1)(a+b)}{2}$$

which simplifies to

$$runsum(a,b) = \frac{b^2-a^2+a+b}{2}$$

add $2a - 2a$ to the numerator:

$$runsum(a,b) = \frac{b^2-a^2+b-1}{2} + \frac{2a}{2}$$

$$runsum(a,b) = T_b - T_a + a$$

$$T_a = \frac{a(a+1)}{2}$$

$$T_b = \frac{b(b+1)}{2}$$

$$T_b - T_a = \frac{b(b+1)}{2} - \frac{a(a+1)}{2}$$

$$= \frac{b^2+b-a^2-a}{2} = \frac{b^2-a^2+b-a}{2}$$

noting the difference of squares pattern, we have

$$T_b - T_a = \frac{(b-a)(b+a+1)}{2}$$

The runsum of integers between a and b inclusive, where $a < b$, is the difference between the triangular numbers ($T_b - T_a$) to which the lower limit, *a,* is added.

•

APPENDIX

Features of Pascal's Triangle

Pascal's Triangle left-justified

Row # ↓ / Col # →		0	1	2	3	4	5	6	7	8	9	10	11	12
0		1	①	②	③									
1		1	1											
2		1	2	1										
3	④	1	3	3	1	$11^3 = 1331$								
4		1	4	6	4	1								
5		1	5	10	10	5	1							
6	⑤	1	6	15	20	15	6	1	Row sum = $2^6 = 64$					
7		1	7	21	35	35	21	7	1					
8	⑥	1	8	28	56	70	56	28	8	1	$(a + b)^8$ expansion coefficients			
9		1	9	36	84	126	126	84	36	9	1	⑧		
10	⑦	1	10	45	120	210	252	210	120	45	10	1		
11		1	11	55	165	330	462	462	330	165	55	11	1	
12		1	12	66	220	495	792	924	792	495	220	66	12	1

1. **Natural numbers.** The "counting" numbers.
2. **Triangular numbers.** The number of dots it takes to make various sized triangles.
3. **Tetrahedral numbers, or triangular pyramidal numbers.** The number of dots needed to make pyramids with triangle bases.
4. **Powers of 11.** The digits form the number of 11 raised to the power of the row number.
5. **Powers of 2.** The sum of the row entries is 2 raised to the power of the row number.
6. **Binomial expansion.** Each column entry is the coefficient of the term of the binomial expansion to the power of the row number.
7. **Perfect squares.** The square of the number in column 1 is the sum of the two adjoining numberrs in column 2.
8. **Combinations** $\binom{\text{Row \#}}{\text{Column \#}} = \binom{m}{n}$ where $0 \leq n \leq m$

INDEX

• • •

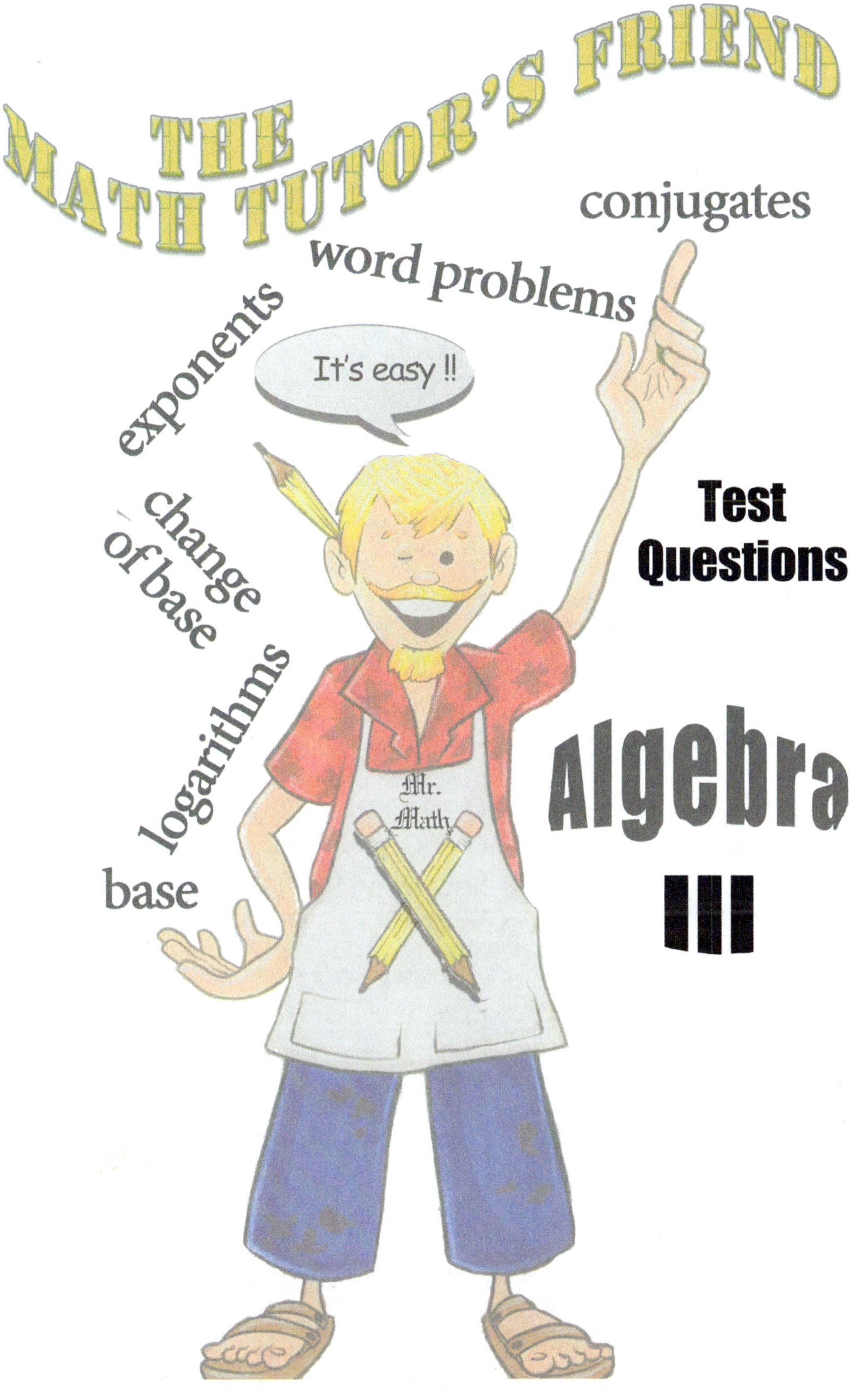
THE MATH TUTOR'S FRIEND
conjugates
word problems
exponents
It's easy !!
Test Questions
change of base
logarithms
Mr. Math
Algebra III
base

QUIZZES

●

• Jack can paint the interior of a closet in 4 hours. Sandy, being more careful, can paint it in 6 hours. If they work together, how long will it take to paint the closet (assuming enough room to do it together)?

A. 6.0 hours
B. 4.2 hours
C. 2.0 hours
D. 2.4 hours
E. None of the above

• • Answer D. The underlying setup is

1 job/4hrs + 1 job/6hrs = 1 job/T hrs: $1/4 + 1/6 = 1/T$; cross-multiplying the left side: $10/24 = 1/T$. Take the reciprocal of both sides: $T = 24/10 = 2.4$ hours

• ♥ •

• Assume a 7% sales tax. What is the price, P, of a taxable item sold for $100.00, tax included?

A. $100.07
B. $107.00
C. $93.00
D. $93.46
E. None of the above

• • Answer D.

$P + 0.07P = \$100.00$

$P(1.07) = \$100.00$

$P = \$100.00/1.07 = \$93.46$

• A theater sold tickets for a show: \$15.00 for adults and \$10.00 for "seniors." 150 tickets sold for \$2,235. How many were "seniors" tickets? (Hint: Let A and S be the number of Adult and Seniors tickets sold, then S = 150 – A).

A. 75
B. 3
C. 15
D. No "senior" tickets sold
E. None of the above

• • Answer B.

The number sold to senior citizens is 150 – A.

Total prices for each group are $15A + 10(150 - A) = 2235$

$15A + 1500 - 10A = 2235$	[Clear the parentheses]
$5A = 2235 - 1500$	[Collect like terms]
$5A = 735$	[Simplify]
$A = 147$	[Divide both sides by 5]
$S = 150 - A = 150 - 147 = 3$	[Number of Sr tickets sold]

• ♥ •

• Twice a number increased by 4 is equal to 8 plus the number. What is the number?

____

• • Answer: 4.

First, identify the unknown number as, say, *n*.

Now write the statement again: $2 \times n + 4 = 8 + n$ from which *n* is easily determined to be 4.

• The radius of the wheels on a car is 18 inches. The car moves forward and the wheels make 1,000 rotations. Assuming there is now skidding, how far did the car travel? Give an exact answer.

A. 5,000 feet
B. 9,425.8 feet
C. $3{,}000\pi$ feet
D. $5{,}000\pi$ feet
E. None of the above.

• • Answer C.

Each full rotation of the wheels moves the car a distance equal to the circumference of the wheel ($2\pi \times 1.5$ feet), so the car will have travelled 1,000 times the circumference: distance traveled = $1{,}000 \times 3\pi$ or $3{,}000\pi$ feet (exact solution)

($1{,}000 \times 3\pi = 3{,}000\pi \approx 9{,}425.8$ feet is an approximate solution.)

• ♥ •

• $x = \log_4 16 =$

A. 2
B. 4
C. 16
D. 32
E. No solution

• • Answer A.

By definition, $4^x = 16$ from which $x = 2$

• Convert 2350% to a decimal and 35% to a fraction

A. 23.5 and 0.35
B. 2,350 and 35/100
C. 23.5 and 7/20
D. 2.35 and 35/100
E. None of the above

• • Answer C.

To convert a % to a decimal remove the "%" symbol and move the decimal point two places to the left. 2,350% becomes 23.5

To convert a % to a fraction remove the "%" symbol and divide by 100.

35% = 35/100 = 7/20 (lowest terms)

• ♥ •

• Combine ($\frac{1}{2}\log(x) - 8\log(y) + 4$) into a single logarithm expression with coefficient of 1. Hint: $\log(1000) = \log(10^3) = 3\log(10) = 3$.

A. $\log(y^8)$
B. $\log(10{,}000\sqrt{x}/y^8)$
C. $\log(\sqrt{x}/y^8)$
D. $\log(x^{1/2})$
E. None of the above.

• • Answer B.

$\frac{1}{2}\log(x) - 8\log(y) + 4$

$\log(x^{1/2}) - \log(y^8) + \log(10^4)$

$= \log(x^{1/2}) - (\log(y^8) - \log(10^4)) = \log(x^{1/2}) - \log(y^8/10^4)$

$= \log[x^{1/2}/(y^8/10^4)] = \log(10{,}000\sqrt{x}/y^8)$

• Simplify $(x^4 - 2^4)/(x^2 + 4)$ to an expression without exponents

A. $(x + 2)$
B. $(x - 2)$
C. $(x + 2)(x - 2)$
D. $2x - 1$
E. None of the above

• • Answer C.

$$(x^4 - 2^4)/(x^2 + 4) = (x^2 - 2^2)(x^2 + 2^2)/(x^2 + 4)$$
$$= (x^2 - 4)(x^2 + 4)/(x^2 + 4)$$
$$= (x^2 - 2^2) = (x - 2)\ (x + 2)$$

• ♥ •

• Write an equivalent expression with no exponents:

$$\ln(x^3\sqrt{(3^2 + 2^4)})$$

A. $\ln(x)$
B. $75\ln(x)$
C. $3\ln(x) + \ln(5)$
D. $15x^2 - \sqrt{25}$
E. None of the above

• • Answer C.

$$\ln(x^3\sqrt{(3^2 + 2^4)}) = \ln(x^3) + \ln\sqrt{25} = 3\ln(x) + \ln(5)$$

• Solve for x

$$f(x) = (x - 2 - \sqrt{3})\,(x - 2 + \sqrt{3}) = 0$$

A. $x = 3$

B. $x = 2 + \sqrt{3}$

C. $x = \sqrt{3}$

D. $x = 2 \pm \sqrt{3}$

E. None of the above

• • Answer D.

Solve by inspection or group to form a "difference of squares" pattern

$(x - 2 - \sqrt{3})\,(x - 2 + \sqrt{3}) = 0$

$[(x - 2) - \sqrt{3}][(x - 2) + \sqrt{3}] = 0$

$(x - 2)^2 - 3 = 0$

$(x - 2)^2 = 3$

$(x - 2) = \pm\sqrt{3}$ $\qquad x = 2 \pm \sqrt{3}$

• ♥ •

• $(2\sqrt{6} - \sqrt{8})\,(2\sqrt{6} + \sqrt{8}) =$

A. 8

B. 12

C. 16

D. 48

E. None of the above

• • Answer C.

This is a "difference of squares" pattern:

$$(2\sqrt{6} - \sqrt{8})\,(2\sqrt{6} + \sqrt{8}) = 4\times 6 - 8 = 24 - 8 = 16$$

• If $\log_2 5 \approx 2.3219$, calculate $x = \log_2 25$.

A. $x \approx 5$
B. $x \approx 2.5$
C. $x \approx 2.321$
D. $x \approx 4.6438$
E. None of the above

• • Answer D.

$$x = \log_2 25 = \log_2 5^2 = 2\log_2 5 \approx 2 \times 2.3219 \,,$$

Therefore, $x \approx 4.6438$

• ♥ •

• $\log_4(0.6) = ?$

A. 2.4
B. $\log_4 3 - \log_4 5$
C. $\log_2(2.4)$
D. 24
E. None of the above

• • Answer B.

$$\log_4(0.6) = \log_4(6/10) = \log_4(3/5) = \log_4 3 - \log_4 5$$

- Simplify the term and rationalize the denominator. $\sqrt{\dfrac{a^5}{b^7c^9}}$

  A. $\dfrac{a^2}{b^3c^4}\sqrt{\dfrac{a}{bc}}$

  B. $\dfrac{x^2}{b^4c^5}$

  C. $\sqrt{abc}$

  D. abc

  E. $\dfrac{a^2\sqrt{abc}}{b^4c^5}$

• • Answer E.

It may be easier to work with the fractional exponent 1/2

$$\left(\frac{a^5}{b^7c^9}\right)^{\frac{1}{2}} = \left(\frac{aa^4}{bb^6cc^8}\right)^{\frac{1}{2}} = \frac{a^2}{b^3c^4}\sqrt{\frac{a}{bc}}$$

$$= \frac{a^2\sqrt{a}}{b^3c^4\sqrt{bc}} \times \frac{\sqrt{bc}}{\sqrt{bc}} = \frac{a^2\sqrt{abc}}{b^4c^5}$$

• ♥ •

- Simplify to an expression without negative exponent or parentheses:

$$(x/y)^{-2}$$

  A. x/y

  B. x/y^2

  C. $(y/x)^2$

  D. y^2/x^2

  E. None of the above

• • Answer D. $(x/y)^{-2} = (y/x)^2 = y^2/x^2$

• Cindy is younger than Joe by 38 years; the sum of their ages is 148. What are their ages?

• • Answer

Cindy is 63 and Joe is 85.
J – C = 38 and J + C = 148.
From the first equation, J = C + 38; substitute for J in the second equation to get

$$38 + 2C = 148$$

$$2C = 148 - 22 = 126$$

$$C = 63 \text{ and } J = C + 22 = 63 + 22 = 85.$$

• ♥ •

•Express the ratio of 35 less than x to 15 more than x.

____

• • Answer:

All the stuff to the left of "*to*" forms the numerator of the ratio and stuff to the right of "*to*" forms the denominator, so the ratio is expressed as

$$(x - 35)/(x + 15).$$

• Write a seven-digit number that is divisible by 6 such that exactly four of the digits are 1.

• • Answer: There are many solutions. To be divisible by 6 a number must be even and also divisible by 3 (sum of digits are divisible by 3)

One solution is to just write four "1s" followed by an even digit: 11112. Now the digits do sum to 6 which is divisible by 3, but we need a seventh digit, so either insert a "0," "3," "6," or "9" somewhere, e.g.,

101112

• ♥ •

• $(12 + i5) \div (3 - i) =$

A. $3.1 + i2.7$

B. $4 - i$

C. $4 + i$

D. $4 - i5$

• • Answer A.

$(12 + i5)/(3 - i) = (12 + i5) \times (3 + i)/[(3 - i)(3 + i)]$

$$= (12 + i5) \times (3 + i)/(9 + 1)$$
$$= (36 + i12 + i15 + 5i^2)/10$$
$$= (36 + i27 - 5)/10 = 3.1 + i2.7$$

• Solve for x: $\ln(x) - \ln(5) = \ln(1)$

A. $x = 5$

B. $x = 1$

C. $x = 0$

D. $x = -1$

E. None of the above

• • Answer A. $\ln(x) - \ln(5) = \ln(1)$

$\ln(x/5) = \ln(1)$

$\ln(x/5) = \ln(5/5)$

$x/5 = 5/5$, therefore $x = 5$.

Also, $\ln(x) - \ln(5) = \ln(1) = 0$ because for any base the log of 1 is zero

$\ln(x) = \ln(5)$ so clearly, $x = 5$.

• ♥ •

• $\log(x) + \log(3x - 5) = \log(2)$

Solve for x.

A. $x = 3$

B. $x = 5$

C. $x = -1/2, 2$ (two solutions)

D. $x = 2$

E. None of the above

• • Answer D. $\log(x) + \log(3x - 5) = \log(2)$

$\log[x(3x - 5]) = \log(2)$

$\log(3x^2 - 5x) = \log(2)$

$3x^2 - 5x = 2$ or $3x^2 - 5x - 2 = 0$

Use the quadratic formula to solve for x:

$x = -1/2, 2$

We are given $\log(x)$ so x cannot be negative

(the logarithm of a negative number is undefined), therefore $x = 2$.

• The diameter of the wheels on a car is 3 feet. The car moves forward and the wheels make 1,000 rotations. To the nearest integer, how far did the car travel, approximately?

A. 1000π feet

B. 10 miles

C. 9425 feet

D. 10,000 feet

E. None of the above

• • Answer C.

Each full rotation of the wheels moves the car a distance equal to the circumference of the wheel, so the car will have travelled 1,000 times the circumference:

$$D = 1{,}000 \times 3\pi \text{ or } 3{,}000\pi \text{ feet (exact solution)}$$

$D = 1{,}000 \times 3\pi = 3{,}000\pi \approx 9{,}425$ feet.

• ♥ •

• What is the area of a circular swimming pool that has a diameter of 20 feet? Give an exact answer.

A. 100π square feet

B. 3142 square feet

C. 1000 square feet

D. 3141.6 square feet

E. An exact answer is not possible

• • Answer A.

The applicable formula is $A = \pi r^2$

An exact answer is possible and will be in terms of π.

(Answers B and C are not measurements and therefore are exact but incorrect)

• 25 is what percent of 300?

A. 25.00%
B. 300.00%
C. 8.33%
D. 12.25%
E. None of the above

• • Answer C.

$$(25/300) \times 100\% = 8.33\% \text{ (to 2 decimal places).}$$

• ♥ •

• What is 23 percent of 49?

A. 11.27%
B. 23.99%
C. 50.00%
D. 49.00%
E. None of the above

• • Answer A.

$$23\% \text{ of } 49 = 0.23 \times 49 = 11.27$$

• A box contains 24 coins. There are twice as many dimes as quarters and three times as many nickels as quarters and twice as many pennies as nickels. How many of each type coin do we have?

A. 4 quarters, 4 dimes, 6 nickels, and 10 pennies
B. 2 quarters, 2 dimes, 8 nickels, and 12 pennies
C. 6 quarters, 6 dimes, 6 nickels, and 6 pennies
D. 2 quarters, 4 dimes, 6 nickels, and 12 pennies
E. None of the above

• • Answer D.

Let q, d, n, and p be the numbers of quarters, dimes, nickels, and pennies, respectively. Reading the problem again, we see that $d = 2q$, $n = 3q$, and $p = 2n$. Just set up an equation in terms of one kind of denomination and then work out the others. Let's put everything in terms of quarters.

There are q quarters
$2q$ dimes because $d = 2q$
$3q$ nickels because $n = 3q$
$2(3q)$ pennies because $p = 2n$ and $n = 3q$.

So, $q + 2q + 3q + 6q = 24$
factoring q: $q(1 + 2 + 3 + 6) = 12q = 24$
from which $q = 24/12 = 2$

There are 2 quarters, 4 dimes, 6 nickels, and 12 pennies which total 24 coins.

• ♥ •

• 32 is 43% of what?

A. 50.00
B. 43.55
C. 47.42
D. 34.32
E. None of the above

• • Answer E.

$0.43x = 32 \quad x = 32/0.43 = 74.42$

• An item costs \$134 and sells for \$175. If the price is increased to \$200, what is the approximate percent of price increase?

A. 25.5%

B. 50.5%

C. 27.4%

D. 14.3%

E. None of the above

• • Answer D.

Find the difference between old and new prices as a percentage of the original

$(200 - 175)/175 = {}^{25}\!/_{175} \approx 0.1429$, a 14.3% increase.

• ♥ •

• Assume a 7% sales tax. What is the price of a taxable item sold for \$100.00, tax included? Round the answer to the nearest cent.

A. \$93.46

B. \$107.10

C. \$93.39

D. \$97.00

E. None of the above

• • Answer A.

Let P be the price without tax.

$P + 0.07P = \$100$. $P = \$100/1.07 \approx \$93.46$.

• Solve for x:

$$(x - 3) + i(y + 4) = 2x + i4$$

A. $x = -3$

B. $x = 4$

C. $x = 0$

D. $x = 13$

E. None of the above

• • Answer A.

Just equate the real parts and solve for x. You can also solve for y (= 0) by equating the imaginary parts of the complex expressions.

• ♥ •

• Solve for x and y:

$$2x + i3y = 28 + i15$$

A. $x = 28; y = 15$

B. $x = 2; y = 3$

C. $x = 28; y = 3$

D. $x = 2; y = 15$

E. None of the above

Answer E.

Just equate the real part and the imaginary parts of the equation to get $x = 14; y = 5$

• The base unit for volume is

A. M
B. T
C. V
D. L
E. None of the above

• • Answer D.
Volume (L^3) is a derived unit which can be factored into base units (i.e., L×L×L)

• ♥ •

• Which of the following can be expressed as a complex number?

A. $\boldsymbol{i}$
B. 34
C. $z = a + \boldsymbol{i}b$
D. All of the above
E. None of the above

• • Answer D.

$\boldsymbol{i} = 0 + \boldsymbol{i}$

$34 = 34 + \boldsymbol{i}0$

• A national budget of \$10 trillion is proposed. This is in the form of a loan which must be repaid. Approximately how long will it take to repay the loan (assume zero interest) if repayment by taxpayers is \$10 thousand a second. Assume one year = 365.25 days.

A. 32 centuries

B. 32 years

C. 3.2 months

D. About one week

E. None of the above

• • Answer B.

$(10^{13}$ dollars×second×hour×day×year)/(10^{4} dollars×3600 seconds×24 hours× 65.25 days)

Dimension analysis shows that dollars, seconds, hours, and days cancel leaving the expression

$$10^{13}\text{ years}/(10^4 \times 3.6 \times 10^3 \times 24 \times 365.25) \approx 31.688 \text{ years.}$$

• ♥ •

• A square has a side length of 30. What is its perimeter?

A. 300
B. 60
C. 90
D. 120
E. None of the above

• • Answer D.

The perimeter is $4 \times 30 = 120$

• Without constructing nor consulting Pascal's Triangle, list the entries in Row 6 of that table. Use of a calculator is permitted.

• • Answer 1 6 15 20 15 6 1

We note that Row 6 has seven entries, that the 1st and 7th entries are 1, the 2nd and 6th entries are 6 (the row number), and that the row is a palindrome (symmetric), so there are only two values to calculate: ${}_6C_2$ and ${}_6C_3$.

$${}_6C_2 = 6!/[(6-2)! \times 2!] = 6\times5\times4!/(4!\times2) = 15$$

Similarly,

$${}_6C_3 = 6!/[(6-3)! \times 3!] = 6\times5\times4\times3!/3!(3\times2\times1) = 20$$

Finally, because of symmetry,

$${}_6C_3 = {}_6C_2 = 15$$

• ♥ •

• Expand $(a+b)^7$

Use the ( ) notation for coefficients of the terms

Answer:

$$(a+b)^7 = \binom{7}{0}a^7 + \binom{7}{1}a^6b + \binom{7}{2}a^5b^2 + \binom{7}{3}a^4b^3 + \binom{7}{4}a^3b^4 + \binom{7}{5}a^2b^5 + \binom{7}{6}ab^6 + \binom{7}{7}b^7$$

Note that the degree of each term is 7

- Convert 4/5 to an Egyptian fraction.

• • First note that 4 is greater than 1/2 of 5, so we can begin with
4/5 =1/2+ x. Solving for x: $x = (8 - 5)/(2\times5) = 3/10$
3/10 = 1/10 + 2/10 = 1/10 + 1/5. So, one of the Egyptian fractions is

$$4/5 = 1/2 + 1/5 + 1/10$$

• ♥ •

- Two oblong numbers associated with 12 are...
  A. 12 and 13
  B. 14 and 15
  C. 131 and 156
  D. 1213 and 1312
  E. None of the above

• • Answer C.

An oblong number, denoted by $\mathbf{O}_n$ is the product of two consecutive integers: $\mathbf{O}_n = n(n + 1)$.

Therefore, for $n = 12$, $\mathbf{O}_{12} = 12\times13 = 156$

and for $(n + 1) = 12$, $\mathbf{O}_{11} = 11\times12 = 131$.

• The sum of consecutive integers 34 through 56, inclusive is

A. S(34, 56)

B. $T_{56} - T_{34}$

C. 1035

D. All of the above

E. None of the above

• • Answer A or C

It is runsum S(34, 56) which is 1035. The difference between the triangular numbers T_{56} and T_{34} is $(T_{56} - T_{34}) = 1596 - 595 = 1001$, to which if we add 34 is equal to 1035, the runsum(34, 56).

• ♥ •

• The difference between triangular number T_{10} and runsum S(1, 10) is...

A. zero

B. $2 \times T_{10}$

C. 9

D. 10

E. None of the above

• • Answer A. They are the same thing, namely, the sum of the consecutive integers 1 through 10, inclusive.

• How can you use Pascal's Triangle to find 2^6 ?

• • Add the entries in Row 6: $1 + 6 + 15 + 20 + 15 + 6 + 1 = 64 = 2^6$.

• ♥ •

• How can you use Pascal's Triangle to find 11^4 ?

• • Copy (concatenate) the entries in Row 4: 1 4 6 4 1 $11^4 = 14641$

• The value of the element in Row 10 and Column 6 of Pascal's Triangle is

____

• • Answer 210.

The value is that of the combination of ten [Row #] things six [Column #] at a time, ${}_{10}C_{6}$.

• ♥ •

THE MATH TUTOR'S FRIEND

INFORMAL REVIEW OF HIGH-SCHOOL MATH TOPICS FOR TUTORS

- 4 -

· Algebra III ·

Printed in Dunstable, United Kingdom

91233123R00078